ServSafe®

Essentials

Third Edition

National Restaurant Association
EDUCATIONAL FOUNDATION

DISCLAIMER

ISBN 1-58280-113-4 (Essentials without Exam—ES3)
ISBN 1-58280-112-6 (Essentials with Exam—ESX3)
ISBN 0-471-52184-1 (Wiley Essentials without Exam—ES3-W)
ISBN 0-471-47803-2 (Wiley Essentials with Exam—ESX3-W)

Printed in the U.S.A.

10 9 8 7 6 5 4 3

Table of Contents

UNIT 3 SANITARY FACILITIES AND PEST MANAGEMENT

A MESSAGE FROM
The National Restaurant Association Educational Foundation

The National Restaurant Association Educational Foundation (NRAEF) is a not-for-profit organization dedicated to fulfilling the educational mission of the National Restaurant Association. Focusing on three key strategies of **risk management, recruitment,** and **retention,** the NRAEF is the restaurant and foodservice industry's premier provider of educational resources, materials, and programs. Sales from all NRAEF products and services benefit the industry by directly supporting the NRAEF's educational initiatives.

Risk management is critical to the success of every restaurant and foodservice operation, particularly the area of food safety. After all, **serving safe food is not an option**—it is an obligation of all restaurant and foodservice professionals. Proper training is one of the best ways to create a culture of food safety within your establishment.

By opening this book, you have made a significant commitment to promoting food safety. You are about to gain knowledge from the **industry standard** in food safety training. The ServSafe® program provides **accurate, up-to-date information** for all levels of employees on all aspects of handling food, from receiving and storing to preparing and serving—information your employees need, to be part of the food safety team.

In the past year, the NRAEF has further enhanced the ServSafe program. Exciting new features include:

▶ **Third editions of *ServSafe Coursebook* and *ServSafe Essentials.*** With increased readability and a new look and format, your best source for food safety training is now even better. Both texts incorporate **the latest information from the 2003 supplement to the 2001 FDA Food Code** in a realistic, applicable manner.

▶ **American National Standards Institute (ANSI)–Conference for Food Protection (CFP) Accreditation.** The ServSafe Food Protection Manager Certification Program has been evaluated and recognized under ANSI–CFP requirements for the development, administration, and security of its examinations. The ServSafe Certification has been deemed valid, reliable, and legally defensible, as well as transferable between jurisdictions.

▶ **Online score access.** You are now able to obtain score information within fifteen days of your examination date. At the time of the examination, ask your instructor for your class identification number, which you will need to retrieve your score at www.nraef.org/classes.

We applaud you for making the commitment to food safety training. The NRAEF challenges you—and all managers—to share your food safety knowledge with your employees and help foster a food safety culture within your operation.

For more information on the NRAEF and its programs, please visit www.nraef.org.

The National Restaurant Association Educational Foundation's

International Food Safety Council®

The International Food Safety Council is a strategic initiative of the National Restaurant Association Educational Foundation created to heighten awareness of the importance of food safety education throughout the restaurant and foodservice industry.

Initiatives

The NRAEF's International Food Safety Council sponsors food safety events and activities to encourage a commitment to food safety education and ensure food safety is a priority.

▶ *Food Safety Illustrated* magazine

▶ Food Safety Summit—Washington, D.C.

▶ National Food Safety Education Month®—September, visit www.nraef.org/nfsem

For more information about the NRAEF's International Food Safety Council, sponsorship opportunities, and initiatives, please call 312.715.1010, ext. 744, or visit the NRAEF's Web site at www.nraef.org/ifsc.

Founding Sponsors
American Egg Board
Cattlemen's Beef Board and
 National Cattlemen's Beef Association
Ecolab, Inc.
FoodHandler Inc.
Heinz North America
San Jamar/Katch All
SYSCO Corporation
Tyson Foods, Inc.

Campaign Sponsors
Bunzl Distribution, Inc.
Cargill, Inc.
Colgate-Palmolive Company
Cooper-Atkins Corporation
Farquharson Enterprises, Ltd.
Jones Dairy Farm
North American Association of Food
 Equipment Manufacturers
Orkin Commercial
PepsiCo Foodservice Division
The Procter & Gamble Company
Produce Marketing Association
UBF Foodsolutions North America
U.S. Foodservice, Inc.

Acknowledgements

The development of *ServSafe Essentials* would not have been possible without the expertise of our many advisors and manuscript reviewers. The NRAEF is pleased to thank the following people for their time, effort, and dedication in making the refinements found in this third edition.

Bruno Abate
Follia Restaurant

Marie-Luise Baehr
Sodexho, Inc.

Harry D'Ercole
Enrico's Italian Dining

Jane Gibson
**Cattlemen's Beef Board and
National Cattlemen's Beef Association**

Steven F. Grover, R.E.H.S.
National Restaurant Association

Bucky Gwartney, Ph.D.
National Cattlemen's Beef Association

Alice M. Heinze
American Egg Board

Todd McAloon
Cargill, Inc.

Anne Munoz-Furlong
Food Allergy and Anaphylaxis Network

Ann Rasor
North American Meat Processors Association

Mary Sandford
Burger King Corporation

SYSCO Corporation

Lacie Thrall
FoodHandler Inc.

Frank Yiannas
Walt Disney World Co.

HOW TO USE
SERVSAFE ESSENTIALS

Suggested below is a plan for studying and retaining the food safety knowledge in this textbook that is vital to keeping your establishment safe.

Beginning Each Section

Prepare for the section by completing the following:

▶ **Review the Learning Objectives.** Located on the front page of each section, the learning objectives identify tasks you should be able to do after finishing the section. They are linked to the essential practices for keeping your establishment safe.

▶ **Test Your Food Safety Knowledge.** Before you begin reading, test your prior knowledge of some of the section's concepts by answering five True or False questions. If you want to explore the concepts behind the questions further, see the page references provided. Answers are located at the back of the section.

Throughout Each Section

Use the following features to help you identify and reinforce the key concepts in the section:

▶ **Concepts.** These topics are important for a thorough understanding of food safety. They are identified before the introduction to each chapter.

▶ **Exhibits.** These are placed throughout each section to visually reinforce the key concepts presented in the text. They are referenced by the section number followed by a letter, and they include charts, photographs, illustrations, and tables.

▶ **Icons.** Three types of icons appear in *ServSafe Essentials*.

   ▶ In Sections 5–10, an icon representing the various stages of the flow of food appears in the left margin. As you read through these sections, you will notice that the highlighted portion of the icon changes according to the area within the flow of food being discussed.

▷ In Section 11, an icon representing the three major categories of content—Sanitary Facilities and Equipment, Cleaning and Sanitizing, and Integrated Pest Management—discussed in the section appears in the left margin. As you read through these sections, you will notice that the highlighted portion of the icon changes according to the content area being discussed.

▷ As part of each "A Case in Point," an icon appears to help illustrate the primary principle addressed in the case study. These icons, shown on the next page, were created by the International Association for Food Protection to provide the restaurant and foodservice industry with easily recognizable symbols that convey specific food safety messages to foodhandlers of all nationalities. For more information on these icons, visit www.foodprotection.org.

▶ **Activities.** Apply what you have learned by completing the various activities throughout the section. Answers are located at the back of the section.

At the End of Each Section

Once you have finished reading and completing the activities throughout the section, see how well you have learned.

▶ **Answer the Multiple-Choice Study Questions.** These questions are designed to test your knowledge of the food safety concepts presented in the section. If you have difficulty answering them, you should review the content further. Answers are located at the back of the section.

Introduction

International Food Safety Icons

Hot Holding

Temperature Danger Zone

Refrigeration/ Cold Holding

Do Not Work If Ill

Cooling

Cross Contamination

Handwashing

No Bare Hand Contact

Wash, Rinse, and Sanitize

Potentially Hazardous Food

Cooking

Icons used with permission from the International Association for Food Protection

HOW TO IMPLEMENT THE FOOD SAFETY PRACTICES LEARNED IN THE SERVSAFE PROGRAM

The ServSafe program will provide you with the essential knowledge to help you keep the food at your establishment safe. It is your responsibility to implement that knowledge in your operation. To do this, you must examine the following aspects of your operation and compare them with your newfound ServSafe knowledge:

▶ Current food safety policies and procedures

▶ Employee training

▶ Your facilities

The steps listed below will help you make the comparison that will take you from where you are today to where you need to go to *consistently* keep food safe in your establishment.

❶ **Evaluate your food safety practices using the *Food Safety Evaluation Checklist* on page A-3.** This checklist identifies the most critical food safety practices that must be followed in every operation. It consists of a series of Yes/No questions that will assist you in identifying opportunities for improvement. Wherever a *No* is checked in this evaluation, you have identified a gap in your practices. These gaps will be the starting point for improving your current food safety program.

❷ **Fill out the *Regulatory Requirements Worksheet* on page A-10.** This worksheet will help you identify areas where your local regulatory requirements and/or company policies differ from the information in *ServSafe Essentials*. It is vital that you comply with your local regulatory requirements. Your company policies should be designed to comply with, or exceed, these requirements. On this worksheet, you will perform a direct comparison of what *ServSafe Essentials* states, what the local jurisdiction requires, and what your company policy states regarding each issue. If you do not have a company policy regarding an issue, this is a gap that must be filled.

❸ **Determine the cause of the gaps identified in Steps 1 and 2.** For example, if your refrigerator is incapable of holding food at 41°F (5°C), this is a gap. There are many things that could have caused this situation, including faulty equipment, a refrigerator door that is opened too frequently, etc. You must explore each of these potential causes to determine the true reason for the gap.

❹ **Address gaps by creating a solution that may include:**

▶ Developing or revising SOPs

   ▶ Training employees on new or revised SOPs

   ▶ Implementing SOPs

▶ Bringing existing equipment up to standard or purchasing new equipment

▶ Training or retraining employees

❺ **Evaluate your solution periodically to ensure it has addressed the gaps identified in Steps 1 and 2.**

Unit 1

The Sanitation Challenge

Providing Safe Food

Inside this section:

▶ The Dangers of Foodborne Illness
▶ Preventing Foodborne Illness

▶ How Food Becomes Unsafe
▶ Keys to Food Safety

After completing this section, you should be able to:

▶ Analyze evidence to determine the presence of foodborne-illness outbreaks.

▶ Recognize risks associated with high-risk populations.

▶ Identify the characteristics of potentially hazardous food.

▶ Recognize a manager's responsibility to provide food safety training to employees.

▶ Identify the need to maintain food safety training records.

▶ Identify the appropriate training tools for teaching food safety.

▶ Ensure all foodservice employees are trained initially and on an ongoing basis.

Apply Your Knowledge	Test Your Food Safety Knowledge

Check to see how much you know about the concepts in this section. Use the page references provided to explore the topic in each question.

❶ **True or False:** A foodborne-illness outbreak is confirmed when two or more people experience the same illness after eating the same food. *(See page 1-4.)*

❷ **True or False:** Preschool-age children may be more likely than adults to become ill from contaminated food. *(See page 1-6.)*

❸ **True or False:** It is the manager's responsibility to teach employees the food safety principles and practices learned in the ServSafe program. *(See page 1-6.)*

❹ **True or False:** Employees only need to receive initial training in food safety. *(See page 1-6.)*

❺ **True or False:** Potentially hazardous food is generally dry, contains protein, and is highly acidic. *(See page 1-7.)*

For answers, please turn to page 1-14.

CONCEPTS

▶ **Foodborne illness:** Disease carried or transmitted to people by food.

▶ **Foodborne-illness outbreak:** Incident in which two or more people experience the same illness after eating the same food.

▶ **Flow of food:** Path food takes from purchasing and receiving, through storing, preparing, cooking, holding, cooling, reheating, and serving.

▶ **FDA Food Code:** Science-based reference for retail food establishments on how to prevent foodborne illness.

▶ **Contamination:** Presence of harmful substances in food. Some food safety hazards occur naturally, while others are introduced by humans or the environment.

▶ **Time-temperature abuse:** Food has been time-temperature abused any time it has been allowed to remain too long at temperatures favorable to the growth of foodborne microorganisms.

▶ **Potentially hazardous food:** Food in which microorganisms can grow rapidly. Potentially hazardous food often has a history of being involved in foodborne-illness outbreaks, has potential for contamination due to methods used to produce and process it, and has characteristics that generally allow microorganisms to thrive. Potentially hazardous food is often moist, contains protein, and has a neutral or slightly acidic pH.

▶ **Cross-contamination:** Occurs when microorganisms are transferred from one surface or food to another.

▶ **Personal hygiene:** Sanitary health habits that include keeping the body, hair, and teeth clean; wearing clean clothes; and washing hands regularly—especially when handling food and beverages.

INTRODUCTION

When diners eat out, they expect safe food, clean surroundings, and well-groomed workers. Overall, the restaurant and foodservice industry does a good job of meeting these demands, but there is still room for improvement.

The risk of foodborne illness impacts the industry. Several factors account for this, and likely include the following:

▶ The emergence of new foodborne pathogens (disease-causing microorganisms)

▶ The importation of food from countries that may not have well-developed, food safety practices

▶ Changes in the composition of food, which may leave fewer natural barriers to the growth of microorganisms

▶ Increases in the purchase of take-out and home meal replacement (HMR) food

► Changing demographics, with an increased number of individuals at high risk for contracting foodborne illness

► Employee turnover rates that make it difficult to manage an effective food safety system

THE DANGERS OF FOODBORNE ILLNESS

A foodborne illness is a disease carried or transmitted to people by food. The Centers for Disease Control and Prevention (CDC) define a foodborne-illness outbreak as an incident in which two or more people experience the same illness after eating the same food. A foodborne illness is confirmed when laboratory analysis shows that a specific food is the source of the illness.

Each year, millions of people are affected by foodborne illness, although the majority of cases are not reported and do not occur at restaurants and foodservice establishments. However, the cases that are reported and investigated help us understand some of the causes of illness, as well as what we, as restaurant and foodservice professionals, can do to control these causes in each of our establishments.

Fortunately, every establishment, no matter how large or small, can take steps to ensure the safety of the food it prepares and serves to its customers.

The Costs of Foodborne Illness

National Restaurant Association figures show that a foodborne-illness outbreak can cost an establishment thousands of dollars. It can even be the reason an establishment is forced to close. *Exhibit 1a* outlines additional costs of a foodborne-illness outbreak.

PREVENTING FOODBORNE ILLNESS

Preventing foodborne illness in your establishment can best be accomplished if you take a comprehensive approach. This includes setting up a food safety management system and training employees to handle food safely. It also includes identifying food that is most likely to become unsafe and the potential hazards that

Exhibit 1a

Costs of a Foodborne Illness to an Establishment

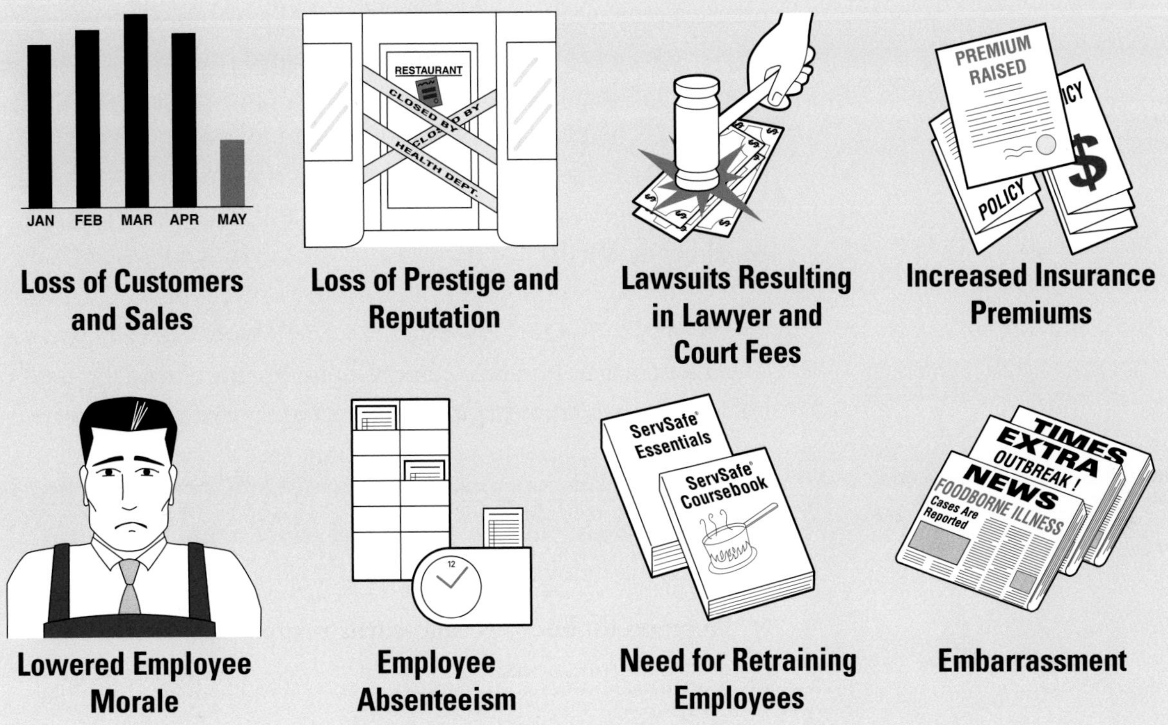

Loss of Customers and Sales

JAN FEB MAR APR MAY

Loss of Prestige and Reputation

RESTAURANT
CLOSED BY
HEALTH DEPT.

Lawsuits Resulting in Lawyer and Court Fees

Increased Insurance Premiums

PREMIUM RAISED

POLICY

Lowered Employee Morale

Employee Absenteeism

Need for Retraining Employees

ServSafe Essentials
ServSafe Coursebook

Embarrassment

TIMES EXTRA
OUTBREAK!
NEWS
FOODBORNE ILLNESS
Cases Are Reported

can contaminate it. Finally, foodborne illnesses can be prevented when high-risk patrons are advised of the risk of consuming raw or undercooked food in your establishment.

Food Safety Systems

A food safety management system will help you prevent foodborne illness by controlling hazards throughout the flow of food. A strong food safety system will incorporate the principles of active managerial control and HACCP.

Active managerial control focuses on establishing policies and procedures to control five common risk factors responsible for foodborne illness. The polices and procedures that an establishment puts in place will be the result of a careful analysis of potential breakdowns related to these five risk factors at each stage in the flow of food.

A HACCP (Hazard Analysis Critical Control Point) system focuses on identifying specific points within the flow of food through the operation that are essential to prevent, eliminate, or reduce a biological, chemical, or physical hazard to safe levels. To be effective, a HACCP system must be based on a plan specific to a facility's menu, customers, equipment, processes, and operation. The HACCP plan is developed following seven principles—sequential steps for building the food safety system. Active managerial control and HACCP will be covered in more detail in Section 10.

Training Employees in Food Safety

As a manager, it is your responsibility to ensure that the food safety principles you learn throughout the ServSafe program are practiced by everyone in your operation. (See *Exhibit 1b.*) All employees must be properly trained in food safety practices that relate to their assigned job tasks. Food safety training should consist of the following:

▶ Programs for both new and current employees

▶ Assessment tools that identify ongoing food safety training needs for employees

▶ A selection of resources that includes books, videos, posters, and technology-based materials to meet your learners' needs

▶ Records documenting that employees have completed training

Populations at High Risk for Foodborne Illness

The demographics of our population show there is an increase in the percentage of people at high risk of contracting a foodborne illness, sometimes with serious consequences. (See *Exhibit 1c.*) They include:

▶ Infants and preschool-age children who have not built up adequate immune systems (the body's defense system against illness)

Exhibit 1b

Training Employees

It is the manager's responsibility to train employees in food safety.

People at High Risk for Foodborne Illness

Young Children

Pregnant Women

Elderly People

People Taking Medication

People Who Are Ill

▶ Pregnant women

▶ Elderly people whose immune systems have weakened with age

▶ People taking certain medications, such as antibiotics and immunosuppressants

▶ People who are ill (those who have recently had major surgery, are organ-transplant recipients, or who have pre-existing or chronic illnesses)

Because these populations are susceptible to foodborne illness, it is of particular concern when they consume potentially hazardous food (or ingredients) that are raw or have not been cooked to the minimum internal temperatures identified in Section 8. In all cases, these high-risk consumers should be advised of any potentially hazardous food (or ingredients) that are raw or not fully cooked. Tell them to consult a physician before regularly consuming this type of food. **Additionally, check with your regulatory agency for specific requirements.**

Food Most Likely to Become Unsafe

Although any food can become contaminated, most foodborne illnesses are transmitted through food in which microorganisms are able to grow rapidly. Such food is classified as potentially hazardous food. This food typically has the following characteristics:

▶ A history of being involved in foodborne-illness outbreaks

▶ A natural potential for contamination due to production and processing methods

▶ Moisture

▶ Contains protein

▶ Neutral or slightly acidic pH levels

The FDA Food Code identifies the food illustrated in *Exhibit 1d* on the next page as potentially hazardous.

Care must be taken to prevent contamination when handling ready-to-eat food, which is any food that is edible without any further washing or cooking. Ready-to-eat food includes washed, whole or cut fruit; vegetables; deli meats; bakery items; spices, sugars, and seasonings; and properly cooked food.

Potential Hazards to Food Safety

Unsafe food usually results from contamination, which is the presence of harmful substances in the food. Some food safety hazards are introduced by humans or by the environment, and some occur naturally.

Food safety hazards are divided into three categories: biological hazards, chemical hazards, and physical hazards.

▶ **Biological hazards** include certain bacteria, viruses, parasites, and fungi, as well as certain plants, mushrooms, and fish that carry harmful toxins.

▶ **Chemical hazards** include pesticides, food additives, preservatives, cleaning supplies, and toxic metals that leach from cookware and equipment.

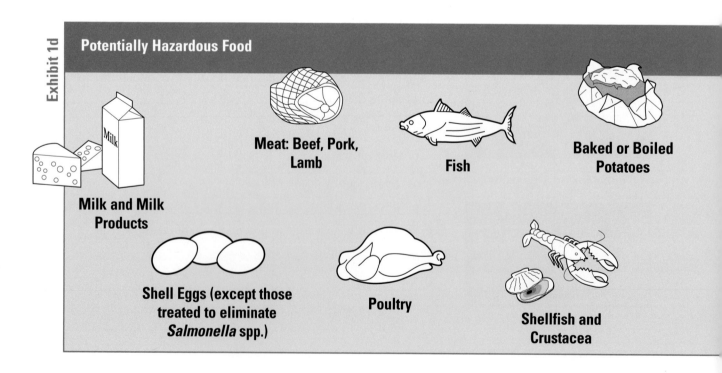

Exhibit 1d

Potentially Hazardous Food

Milk and Milk Products

Meat: Beef, Pork, Lamb

Fish

Baked or Boiled Potatoes

Shell Eggs (except those treated to eliminate *Salmonella* spp.)

Poultry

Shellfish and Crustacea

▶ **Physical hazards** consist of foreign objects that accidentally get into the food, such as hair, dirt, metal staples, and broken glass, as well as naturally occurring objects, such as bones in fillets.

By far, biological hazards pose the greatest threat to food safety. Disease-causing microorganisms are responsible for the majority of foodborne-illness outbreaks.

HOW FOOD BECOMES UNSAFE

The CDC have identified common factors that are responsible for foodborne illness. These include:

▶ Purchasing food from unsafe sources

▶ Failing to cook food adequately

▶ Holding food at improper temperatures

▶ Using contaminated equipment

▶ Poor personal hygiene

Each of these common factors—with the exception of purchasing food from unsafe sources—is related to time-temperature abuse, cross-contamination, or poor personal hygiene. Reported cases of foodborne illness usually involve multiple causes. A well-designed, food safety system will control all of these factors.

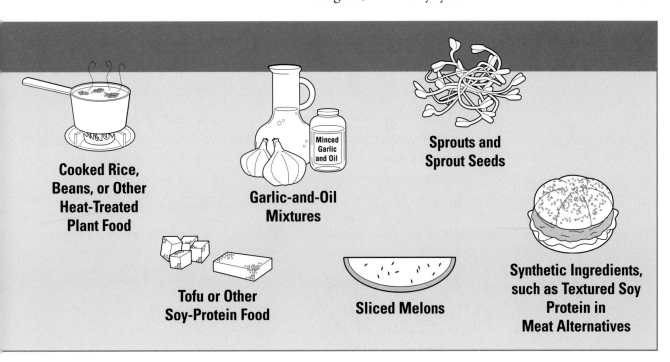

Cooked Rice, Beans, or Other Heat-Treated Plant Food

Garlic-and-Oil Mixtures

Sprouts and Sprout Seeds

Tofu or Other Soy-Protein Food

Sliced Melons

Synthetic Ingredients, such as Textured Soy Protein in Meat Alternatives

Time-Temperature Abuse: Food has been time-temperature abused any time it has been allowed to remain too long at temperatures favorable to the growth of foodborne microorganisms. A foodborne illness can result if food is time-temperature abused in the following manner:

► It is not held or stored at required temperatures.

► It is not cooked or reheated to temperatures that kill microorganisms.

► It is not cooled properly.

► It is prepared a day or more in advance.

Cross-Contamination: Cross-contamination occurs when microorganisms are transferred from one surface or food to another. A foodborne illness can result if cross-contamination is allowed to occur in any of the following ways:

► Raw contaminated ingredients are added to food that receives no further cooking.

► Food-contact surfaces are not cleaned and sanitized before touching cooked or ready-to-eat food.

► Raw food is allowed to touch or drip fluids onto cooked or ready-to-eat food.

► A foodhandler touches contaminated (usually raw) food and then touches cooked or ready-to-eat food.

► Contaminated cleaning cloths are not cleaned and sanitized before being used on other food-contact surfaces.

Poor Personal Hygiene: Individuals with poor personal hygiene can offend customers, contaminate food or food-contact surfaces, and cause illnesses. A foodborne illness can result if employees do any the following:

► Fail to wash their hands properly after using the restroom or whenever they become contaminated

► Cough or sneeze on food

► Touch or scratch sores, cuts, or boils, and then touch food they are handling

► Come to work while sick

Apply Your Knowledge	Potentially Hazardous or Not?
Place an **X** next to each item that is potentially hazardous.	

__ ❶ Raw carrots   __ ❻ Lettuce   __ ⓫ Limes

X ❷ Sliced melons   __ ❼ Bananas   X ⓬ Shell eggs

X ❸ Bean sprouts   __ ❽ Flour   X ⓭ Soy burger

X ❹ Baked potatoes   __ ❾ Dry rice   X ⓮ Cheese

__ ❺ Soda crackers   X ❿ Tofu   __ ⓯ Bread

For answers, please turn to page 1-14.

Apply Your Knowledge	A Case in Point

❶ What did André do right?

❷ What did he do wrong?

❸ What should he do differently in the future?

For answers, please turn to page 1-14.

André, a restaurant manager, has been on the job for nine months at The Charter Café. In his first two weeks, ten new employees joined his team of twenty-five. To his credit, most of the thirty-five employees are still with the establishment.

André knew from his management training program that food safety training was important to the success of the restaurant. With that in mind, he made sure all of his employees went through a four-hour food safety training session that he presented himself. André tried his best to impart the knowledge he learned in his training to his employees. He even had them take a quiz that he created.

After five months on the job, André found himself responding to a foodborne-illness complaint and a failing health inspection score. He was confused about what could have gone wrong because all of his employees had been trained in food safety.

KEYS TO FOOD SAFETY

The keys to food safety lie in controlling time and temperature throughout the flow of food, practicing good personal hygiene, and preventing cross-contamination. It is important to establish standard operating procedures that focus on these areas. The ServSafe program will provide you with the knowledge to properly design these procedures.

SUMMARY

Foodborne illness is a major concern to the restaurant and foodservice industry. A foodborne illness is a disease carried or transmitted to people by food. A foodborne-illness outbreak involves two or more people experiencing the same illness after eating the same food. Some segments of the population more susceptible to foodborne illnesses than others are referred to as high-risk populations.

Some food has a history of involvement in foodborne-illness outbreaks. This is called potentially hazardous food. Typically having a natural potential for contamination due to production and processing methods, potentially hazardous food is often moist, contains protein, and has a neutral or slightly acidic pH.

An incident of foodborne illness can be very expensive for an establishment, including legal liability, damage to reputation, and other related factors. However, a well-designed, food safety system protects your customers, your employees, and your reputation. It will incorporate the principles of active managerial control and HACCP. Active managerial control focuses on establishing policies and procedures to control five common risk factors responsible for foodborne illness. A HACCP system focuses on identifying specific points within the flow of food through the operation that are essential to prevent, eliminate, or reduce a biological, chemical, or physical hazard to safe levels.

Key practices for ensuring food safety include controlling time and temperature, practicing strict personal hygiene, and preventing cross-contamination. Also receive, store, prepare, cook, hold, cool, reheat, and serve food using methods that maintain its safety. Finally, you must carefully train, monitor, and reinforce food safety principles in your establishment.

Apply Your Knowledge   Multiple-Choice Study Questions

Use these questions to test your knowledge of the concepts presented in this section.

1. Why do elderly people have a higher risk of contracting a foodborne illness?
   A. They are more likely to spend time in a hospital.
   B. Their immune systems are likely to have weakened with age.
   C. Their allergic reactions to chemicals used in food production might be greater than those of younger people.
   D. They are likely to have diminished appetites.

2. Which type of food would be the most likely to cause a foodborne illness?
   A. Tomato juice   C. Whole wheat flour
   B. Cooked rice   D. Dry powdered milk

3. Which of the following is *not* a common characteristic of potentially hazardous food?
   A. They are moist.
   B. They are dry.
   C. They are neutral or slightly acidic.
   D. They contain protein.

4. In order for a foodborne illness to be considered an "outbreak," how many people must experience the illness after eating the same food?
   A. 1   B. 2   C. 10   D. 20

5. Food safety training should consist of all of the following *except*
   A. programs for new and current employees.
   B. assessment tools that identify training needs.
   C. records that document that training has occurred.
   D. methods for dealing with customers' complaints.

6. Tools used for food safety training should
   A. meet the needs of your learners.
   B. be inexpensive.
   C. be deliverable in five minutes.
   D. resemble a video game.

For answers, please turn to page 1-14.

Apply Your Knowledge   Answers

Page	Activity

1-2   Test Your Food Safety Knowledge

1. False   2. True   3. True   4. False   5. False

1-11   Potentially Hazardous or Not?

1. No	6. No	11. No
2. Yes	7. No	12. Yes
3. Yes	8. No	13. Yes
4. Yes	9. No	14. Yes
5. No	10. Yes	15. No

1-11   A Case in Point

❶ André tried to ensure that all of his employees went through initial food safety training. ❷ He failed to periodically retrain his employees on food safety practices. Without the refresher training, his employees were likely to forget proper food safety practices and procedures, which led to the lower inspection scores and a potential foodborne illness. ❸ In the future, André should ensure that employees receive initial food safety training and refresher training on an ongoing basis. He should also create or utilize a tool that assesses his employees knowledge and compare it to what they should know. Any gap that is identified should be filled using follow-up training.

1-13   Multiple-Choice Study Questions

1. B
2. B
3. B
4. B
5. D
6. A

Apply Your Knowledge   **Notes**

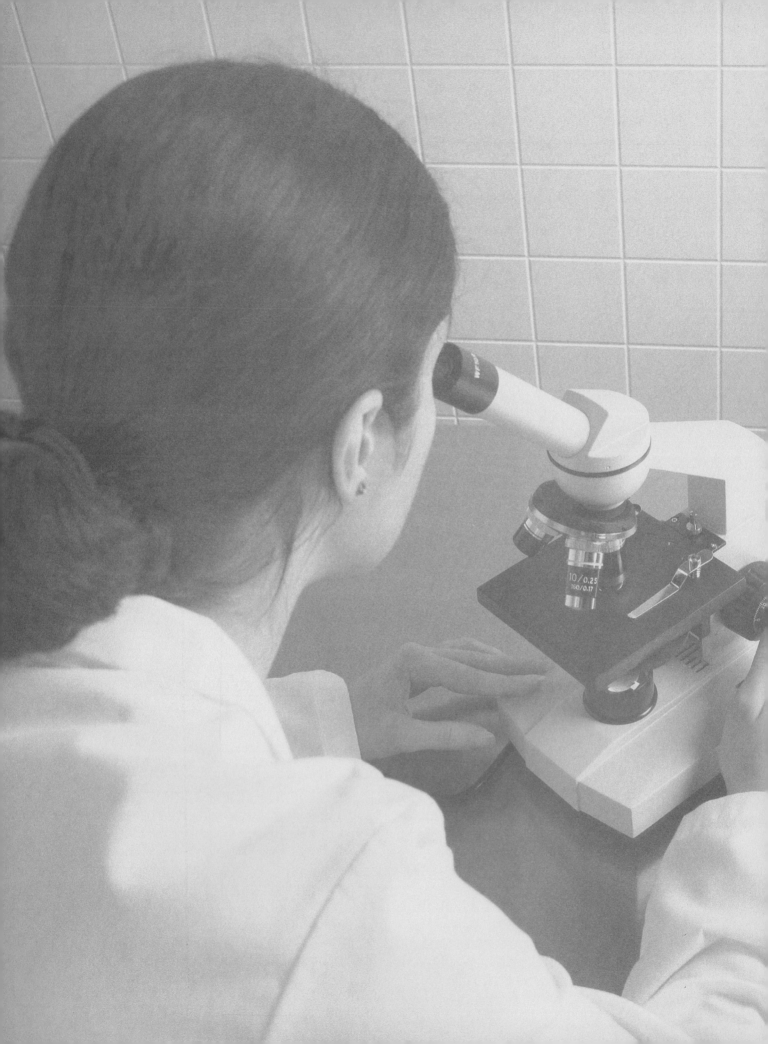

The Microworld

Inside this section:

- Microbial Contaminants
- Bacteria
- Viruses
- Parasites

- Fungi
- Foodborne Infection vs. Foodborne Intoxication

After completing this section, you should be able to:

- Identify factors that affect the growth of foodborne pathogens (FAT TOM).
- Differentiate between foodborne intoxication, infections, and toxin-mediated infections.

- Identify major foodborne illnesses and their symptoms.
- Identify characteristics of major foodborne pathogens including sources, food involved in outbreaks, and methods of prevention.

Apply Your Knowledge	Test Your Food Safety Knowledge

Check to see how much you know about the concepts in this section. Use the page references provided to explore the topic in each question.

① True or False: *Anisakis simplex* is often found in raw seafood. *(See page 2-16.)*

② True or False: A foodborne intoxication occurs when a person eats food containing pathogens, which then grow in the intestines and cause illness. *(See page 2-20.)*

③ True or False: A person with listeriosis may experience bloody diarrhea. *(See page 2-10.)*

④ True or False: Cooling rice properly can help prevent an outbreak of *Bacillus cereus* Gastroenteritis. *(See page 2-11.)*

⑤ True or False: Highly acidic food typically does not support the growth of foodborne microorganisms. *(See page 2-7.)*

For answers, please turn to page 2-24.

CONCEPTS

▶ **Microorganism:** Small, living organism that can be seen only with the aid of a microscope. The four types of microorganisms that can contaminate food and cause foodborne illness are bacteria, viruses, parasites, and fungi.

▶ **Pathogens:** Disease-causing microorganisms.

▶ **Bacteria:** Single-celled living microorganisms that can cause food spoilage and foodborne illness. Some form spores and can survive freezing and very high temperatures.

▶ **Virus:** The smallest of the microbial food contaminants, viruses rely on a living host to reproduce. They usually contaminate food through a foodhandler's improper personal hygiene. Some might survive freezing and cooking temperatures.

▶ **Parasite:** Organism that needs to live in a host organism to survive. Parasites can live inside many animals that humans use for food, such as cows, chickens, pigs, and fish. Proper cooking and freezing will kill parasites. Avoiding cross-contamination and practicing proper handwashing can also prevent foodborne illness caused by parasites.

▶ **Fungi:** Fungi range in size from microscopic, single-celled organisms to very large, multicellular organisms. Fungi most often cause food spoilage. Mold, yeast, and mushrooms are examples of fungi.

▶ **pH:** Measure of a food's acidity or alkalinity. The pH scale ranges from 0 to 14.0. A pH above 7.0 is alkaline, while a pH below 7.0 is acidic. A pH of 7.0 is neutral. Pathogenic bacteria grow well in food with a pH between 4.6 and 7.5 (slightly acidic to neutral).

▶ **Spore:** Alternative form for some bacteria. The spore's thick walls protect the bacteria from adverse conditions, such as high and low temperatures, low moisture, and high acidity. The spore is capable of turning back into a vegetative microorganism when conditions again become favorable for growth.

▶ **Vegetative microorganism:** Bacteria in the process of reproduction. Bacteria reproduce by splitting in two. As long as conditions are favorable, bacteria can grow and multiply very rapidly, doubling their number as often as every twenty minutes.

▶ **FAT TOM:** Acronym for the conditions needed by most foodborne microorganisms to grow: Food, Acidity, Temperature, Time, Oxygen, Moisture.

▶ **Temperature danger zone:** The temperature range between 41°F and 135°F (5°C and 57°C) within which most foodborne bacteria grow and reproduce.

▶ **Water activity:** Amount of moisture available in food for microorganisms to grow. It is measured on a scale from 0–1.0 with water having a water activity (a_w) of 1.0. Potentially hazardous food has water activity values of .85 or above.

▶ **Mold:** Type of fungus that causes food spoilage. Some molds produce toxins that can cause foodborne illness.

▶ **Yeast:** Type of fungus that causes food spoilage.

▶ **Foodborne infection:** Result of a person eating food containing pathogens, which then grow in the intestines and cause illness. Typically, symptoms of a foodborne infection do not appear immediately.

▶ **Foodborne intoxication:** Result of a person eating food containing toxins (poisons) that cause an illness. The toxins may have been produced by pathogens found on the food or may be the result of a chemical contamination. The toxins might also be a natural part of the plant or animal consumed. Typically, symptoms of foodborne intoxication appear quickly, within a few hours.

▶ **Fooodborne toxin-mediated infection:** Result of a person eating food containing pathogens, which then produce illness-causing toxins in the intestines.

INTRODUCTION

In the previous section, you learned that foodborne microorganisms pose the greatest threat to food safety, and that disease-causing microorganisms are responsible for the majority of foodborne-illness outbreaks. In this section, you will learn about the microorganisms that cause foodborne illness, as well as conditions they require in order to grow. When you understand these conditions, you will begin to see how the growth of microorganisms can be controlled, a topic that will be covered in greater detail in later sections.

Microorganisms are small, living beings that can only be seen with a microscope. While not all microorganisms cause disease, some do. These are called pathogens. Eating food contaminated with foodborne pathogens, or their toxins, is the leading cause of foodborne illness.

MICROBIAL CONTAMINANTS

There are four types of microorganisms that can contaminate food and cause foodborne illness: bacteria, viruses, parasites, and fungi.

These microorganisms can be arranged into two groups: spoilage microorganisms and pathogens. Mold is a spoilage microorganism. While its appearance, smell, and taste is not very appetizing, it typically does not cause illness. Pathogens, like *Salmonella* spp. and the hepatitis A virus, cause some form of illness when ingested. Unlike spoilage microorganisms, pathogens cannot be seen, smelled, or tasted in food.

BACTERIA

Of all microorganisms, bacteria are of greatest concern to the manager. Knowing what bacteria are and understanding the environment in which they grow is the first step in controlling them.

Basic Characteristics of Bacteria that Cause Foodborne Illness

Bacteria that cause foodborne illness have some basic characteristics:

► They are living, single-celled organisms.

► They may be carried by a variety of means: food, water, soil, humans, or insects.

► Under favorable conditions, they can reproduce very rapidly.

► Some can survive freezing.

► Some turn into spores, a change that protects the bacteria from unfavorable conditions.

► Some can cause food spoilage; others can cause illness.

► Some cause illness by producing toxins as they multiply, die, and break down. These toxins are not typically destroyed by cooking.

Vegetative Stages and Spore Formation

Although vegetative bacteria may survive low—and even freezing—temperatures, they can be killed by high temperatures. Some types of bacteria, however, have the ability to change into a different form, called a spore. The spore's thick wall protects the bacteria against unfavorable conditions, such as high or low temperature, low moisture, and high acidity.

While a spore cannot reproduce, it is capable of turning back into a vegetative organism when conditions again become favorable for growth. For example, bacteria in food may form spores when exposed to freezer temperatures, allowing the bacteria to survive. As the food thaws and conditions improve, spores can turn back into vegetative cells and begin to grow in the food.

Since spores are so difficult to destroy, it is important to cook, cool, and reheat food properly.

FAT TOM: What Microorganisms Need to Grow

Conditions favoring the growth of most foodborne microorganisms can be remembered by the acronym FAT TOM. (See *Exhibit 2a.*) Each of these conditions for growth will be explained in more detail in the next several paragraphs.

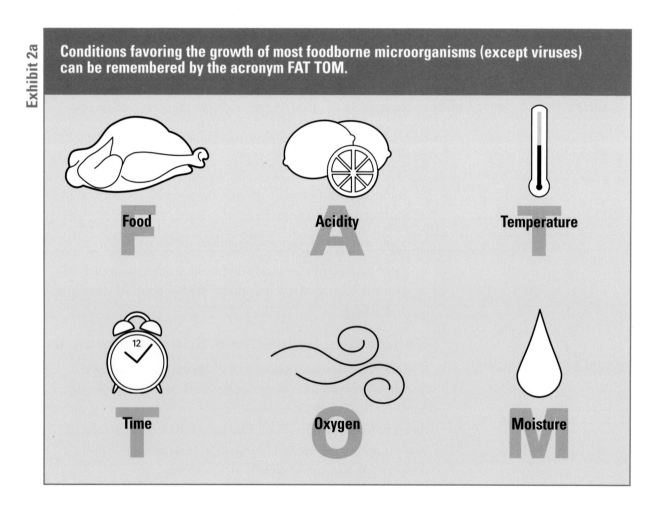

Exhibit 2a

Conditions favoring the growth of most foodborne microorganisms (except viruses) can be remembered by the acronym FAT TOM.

Food

Acidity

Temperature

Time

Oxygen

Moisture

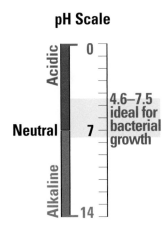

pH Scale

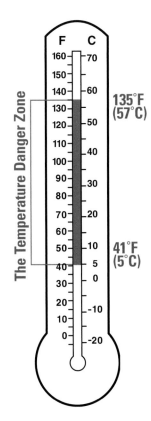

The Temperature Danger Zone

Food

▶ To grow, foodborne microorganisms need nutrients, specifically proteins and carbohydrates. These proteins are commonly found in potentially hazardous food items, such as meat, poultry, dairy products, and eggs. *(See page 1-7.)*

Acidity

▶ Foodborne microorganisms typically do not grow in alkaline or highly acidic foods, such as crackers or lemons. Pathogenic bacteria grow best in food that is slightly acidic or neutral (approximate pH of 4.6 to 7.5), which includes most of the food we eat.

Temperature

▶ Most foodborne microorganisms grow well between the temperatures of 41°F and 135°F (5°C and 57°C). This range is known as the temperature danger zone.

▶ Exposing microorganisms to temperatures outside the danger zone does not necessarily kill them. Refrigeration temperatures, for example, may only slow their growth. Some bacteria grow at refrigeration temperatures. Food must be handled very carefully when it is thawed, cooked, cooled, and reheated since it can be exposed to the temperature danger zone during these times.

Time

▶ Foodborne microorganisms need sufficient time to grow. Bacteria can double their population every twenty minutes.

▶ If potentially hazardous food remains in the temperature danger zone for four hours or more, pathogenic microorganisms can grow to levels high enough to make someone ill. (See *Exhibit 2b* on the next page.)

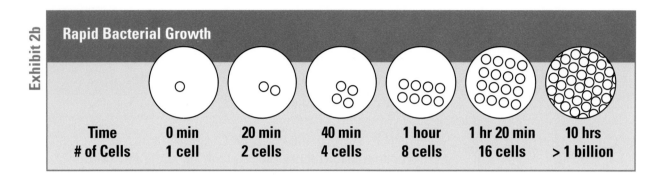

Exhibit 2b

Rapid Bacterial Growth

| Time
# of Cells | 0 min
1 cell | 20 min
2 cells | 40 min
4 cells | 1 hour
8 cells | 1 hr 20 min
16 cells | 10 hrs
> 1 billion |

Oxygen

▶ Some pathogens require oxygen to grow while others grow when oxygen is absent. Pathogens that grow without oxygen can occur in cooked rice, untreated garlic and oil mixtures, and foil-wrapped baked potatoes that have been temperature abused.

Moisture

▶ Because most foodborne microorganisms require water to grow, they grow well in moist food.

▶ The amount of moisture available in a food for microorganisms to grow is called its water activity (a_w). It is measured on a scale from 0 to 1.0, with water having a water activity of 1.0.

▶ Potentially hazardous food items typically have a water activity of .85 or higher.

Controlling the Growth of Microorganisms

FAT TOM is the key to controlling the growth of microorganisms in food, since denying any one of these requirements can prevent growth. Food processors use several methods to keep microorganisms from growing, including:

▶ Adding lactic or citric acid to food to make it more acidic

▶ Adding sugar, salt, alcohol, or acid to a food to lower its water activity

▶ Using vacuum packaging to deny oxygen

While these prevention methods may not be practical for your establishment, there are two important requirements for growth that you can control—time and temperature. Remember that microorganisms grow well at temperatures between 41°F (5°C) and 135°F (57°C). Move food out of this temperature range by cooking it to the proper temperature, freezing it, or by refrigerating it at 41°F (5°C) or lower. In addition, foodborne microorganisms need sufficient time to grow. When preparing food, limit the amount of time it spends in the temperature danger zone and prepare it in small batches, as close to service as possible.

Information about major foodborne illnesses caused by bacteria, as well as ways to prevent them, is presented in *Exhibit 2c* beginning on page 2-10. *(Biological toxins are discussed in Section 3.)*

Apply Your Knowledge	What I Need to Grow!
Place an **X** next to each condition that typically supports the growth of microorganisms.	___ ❶ Food that is high in fat content ___ ❷ Protein food source ___ ❸ pH of 9.0 ___ ❹ Temperature of 155°F (68°C) or higher ___ ❺ Dry environment **For answers, please turn to page 2-24.**

Exhibit 2c

Major Foodborne Illnesses Caused by Bacteria			
Foodborne Illness	**Salmonellosis (nontyphoid)**	**Shigellosis (bacillary dysentery)**	**Listeriosis**
Bacteria	*Salmonella* spp.	*Shigella* spp.	*Listeria monocytogenes*
Symptoms	Nausea, vomiting, abdominal cramps, headache, fever, and diarrhea; may cause severe dehydration in infants and elderly	Diarrhea (may be bloody), abdominal pain, fever, nausea, cramps, vomiting, chills, fatigue, dehydration	Fever and diarrhea are common in individuals who are not immuno-compromised; septicemia, meningitis, encephalitis may result in those who are immuno-compromised (elderly, pregnant women, newborns); may result in stillbirth or abortion of fetuses
Source	Water, soil, insects, domestic and wild animals, and the human intestinal tract; widespread in poultry and swine	Human intestinal tract; flies; frequently found in water polluted by feces	Soil, water, plants; cold damp environments; humans; domestic and wild animals
Food Involved In Outbreaks	Raw poultry and poultry salads; raw meat and meat products; fish; shrimp; milk and dairy products; shell eggs and egg products, such as improperly cooked custards, sauces, and pastry creams; tofu and other protein foods; sliced melons, sliced tomatoes, raw sprouts, and other fresh produce	Salads (potato, tuna, shrimp, chicken, and macaroni); lettuce; raw vegetables; milk and dairy products; poultry	Unpasteurized milk and soft cheeses; raw vegetables; poultry and meat; seafood and seafood products; prepared and chilled ready-to-eat food (e.g., soft cheese, deli foods, pâté, hot dogs)
Preventive Measures	Thoroughly cook poultry to at least 165°F (74°C) for at least 15 seconds and cook other food to required minimum internal temperatures; avoid cross-contamination; properly refrigerate food; properly cool cooked meat and meat products; properly handle and cook eggs; ensure that employees practice good personal hygiene	Ensure that employees practice good personal hygiene when handling ready-to-eat food; avoid cross-contamination; use sanitary food and water sources; control flies; properly cool food	Use only pasteurized milk and dairy products; cook food to required minimum internal temperatures; avoid cross-contamination; clean and sanitize surfaces; thoroughly wash raw vegetables

Major Foodborne Illnesses Caused by Bacteria *continued*		
Foodborne Illness	*Staphylococcal* Gastroenteritis	*Clostridium perfringens* Gastroenteritis	*Bacillus cereus* Gastroenteritis
Bacteria	*Staphylococcus aureus*	*Clostridium perfringens*	*Bacillus cereus*
Symptoms	Vomiting/retching, nausea, diarrhea, abdominal cramps; in severe cases—headache, muscle cramping, changes in blood pressure and pulse rate	Abdominal pain and cramping, diarrhea, nausea (fever, headache, and vomiting usually absent)	Vomiting and nausea, sometimes abdominal cramps or diarrhea (emetic); watery diarrhea, abdominal cramps, pain, nausea (diarrheal)
Source	Humans: nose, skin, hair, throat, and infected sores; animals	Humans and domestic animals (intestinal tracts), soil	Soil and dust; cereal crops
Food Involved In Outbreaks	Reheated or improperly hot-held ready-to-eat food; meat and meat products; poultry, egg products, and other protein food; sandwiches, milk and dairy products; cream-filled pastries; salads (egg, tuna, chicken, potato, and macaroni)	Meat, meat dishes such as stew and gravy, poultry, beans that have been temperature abused	Rice products; starchy food (pasta, potatoes, and cheese products); food mixtures, such as sauces, puddings, soups, casseroles, pastries, salads, and dairy products (emetic); meats; milk and dairy products, vegetables; and fish (diarrheal)
Preventive Measures	Avoid contamination of food from unwashed bare hands; practice good personal hygiene; exclude employees with skin infections from foodhandling and preparation tasks; properly refrigerate food; rapidly cool prepared food	Use careful time and temperature control when holding, cooling, and reheating cooked food	Practice careful time and temperature control when holding, cooling, and reheating cooked food; cook food to required minimum internal temperatures

Major Foodborne Illnesses Caused by Bacteria *continued*

Foodborne Illness	Botulism	Campylobacteriosis	Hemorrhagic Colitis
Bacteria	*Clostridium botulinum*	*Campylobacter jejuni*	Shiga toxin-producing *Escherichia coli*, including O157:H7 and O157:NM
Symptoms	Fatigue, weakness, vertigo followed by blurred or double vision, difficulty speaking and swallowing, dry mouth; eventually leading to paralysis and death	Diarrhea (watery or bloody); fever and nausea; abdominal pain, headache, and muscle pain	Diarrhea (watery, may become bloody); severe abdominal cramps and pain, vomiting, mild or no fever; may cause kidney failure in the very young; symptoms more severe in the immuno-compromised
Source	Present on almost all food of either animal or vegetable origin; soil; water	Poultry and other animals; unpasteurized milk; unchlorinated water	Animals; particularly found in the intestinal tracts of cattle and humans; raw unpasteurized milk
Food Involved In Outbreaks	Food that was under-processed or temperature abused in storage, improperly canned foods, untreated garlic-and-oil mixtures, temperature-abused sautéed onions in butter, leftover baked potatoes, stews, meat/poultry loaves; risk for MAP and *sous vide* products	Unpasteurized milk and dairy products, raw poultry, nonchlorinated or fecal-contaminated water	Raw and undercooked ground beef, unpasteurized milk and apple cider/juice, beef, improperly cured dry salami, lettuce, nonchlorinated water, alfalfa sprouts
Preventive Measures	Do not serve home-canned products; use careful time and temperature control for all bulky, thick foods, purchase only acidified garlic-and-oil mixtures; sauté onions to order or hold them properly; properly cool leftovers	Thoroughly cook food, especially poultry, to required minimum internal temperatures; use pasteurized milk and treated water; avoid cross-contamination	Thoroughly cook ground beef to at least 155°F (68°C) for 15 seconds; avoid cross-contamination, practice good personal hygiene, use only pasteurized milk, dairy products, and juices

Major Foodborne Illnesses Caused by Bacteria *continued*

Foodborne Illness	*Vibrio Parahaemolyticus* Gastroenteritis	*Vibrio vulnificus* Primary Septicemia	Yersiniosis
Bacteria	*Vibrio parahaemolyticus*	*Vibrio vulnificus*	*Yersinia enterocolitica*
Symptoms	Diarrhea, abdominal cramps, nausea, vomiting, headache	Fever, chills, nausea, hypotension, skin lesions may develop	Vary by age group, but diarrhea is common; symptoms may mimic appendicitis
Source	Crabs, clams, oysters, shrimp, lobster, scallops	Raw oysters, particularly those harvested during warmer months; clams, crabs	Domestic animals, soil, water
Food Involved In Outbreaks	Raw or partially cooked oysters, raw or partially cooked shellfish (clams and mussels); cross-contaminated crabs, lobster, shrimp	Raw or partially cooked oysters	Contaminated pasteurized milk, raw unpasteurized milk; tofu; nonchlorinated water; meat (pork, beef, lamb); oysters; fish
Preventive Measures	Tell high-risk populations to consult a physician before regularly consuming raw or partially cooked oysters; purchase seafood from approved suppliers; avoid cross-contamination; maintain time and temperature control	Tell high-risk populations to consult a physician before regularly consuming raw or partially cooked oysters; purchase seafood from approved suppliers; avoid cross-contamination; maintain time and temperature control	Use only pasteurized milk; minimize cross-contamination; thoroughly cook food to required minimum internal temperatures; ensure that utensils and equipment are properly sanitized; use only sanitary, chlorinated water supplies

VIRUSES

Viruses are the smallest of the microbial contaminants. While a virus cannot reproduce outside a living cell, once inside a human cell, it will produce more viruses. Viruses are responsible for several foodborne illnesses, such as hepatitis A.

Basic Characteristics of Viruses

Viruses share some basic characteristics.

▶ Unlike bacteria, they rely on a living cell to reproduce.

▶ They are not complete cells.

▶ Unlike bacteria, they do not reproduce in food.

▶ Some may survive freezing and cooking.

▶ They can be transmitted from person to person, from people to food, and from people to food-contact surfaces.

▶ They usually contaminate food through a foodhandler's improper personal hygiene.

▶ They can contaminate both food and water supplies.

Practicing good personal hygiene is an important way to prevent the contamination of food by foodborne viruses. It is especially important to minimize bare-hand contact with ready-to-eat food. Information about major foodborne viruses, as well as ways to prevent them, is provided in *Exhibit 2d.*

PARASITES

Parasites share some basic characteristics.

▶ They are living organisms that need a host to survive.

▶ They grow naturally in many animals—such as pigs, cats, rodents, and fish—and can be transmitted to humans.

▶ Most are very small, often microscopic, but larger than bacteria.

▶ They pose hazards to both food and water.

Information about major foodborne illnesses caused by parasites, as well as ways to prevent them, is presented in *Exhibit 2e* on page 2-16.

Exhibit 2d

Major Foodborne Illnesses Caused by Viruses

Foodborne Illness	Hepatitis A	Norovirus Gastroenteritis	Rotavirus Gastroenteritis
Virus	*Hepatovirus* or hepatitis A virus	Norovirus (formerly called Norwalk virus)	Rotavirus
Symptoms	Sudden onset of fever; fatigue, nausea, loss of appetite, vomiting, abdominal pain, and jaundice after several days; children often exhibit no symptoms	Nausea, vomiting (more common in children), watery diarrhea with abdominal cramps, mild fever	Vomiting, watery diarrhea, abdominal pain, and mild fever (illness more common in children than adults)
Source	Human intestinal tract; feces-contaminated water	Human intestinal tract and feces-contaminated water	Human intestinal tract; feces-contaminated water
Food Involved In Outbreaks	Shellfish; salads; cross-contaminated deli meats and sandwiches; fruit and fruit juices; milk and milk products; vegetables; any food that will not receive a further heat treatment; water and ice	Ready-to-eat food including salads, sandwiches, and bakery products; liquid items such as salad dressing or cake icing; oysters from contaminated waters; contaminated raspberries; contaminated well water	Water and ice, raw and ready-to-eat food (salads, fruit), contaminated water
Preventive Measures	Obtain shellfish from approved sources; ensure foodhandlers practice good personal hygiene; prevent cross-contamination from hands; clean and sanitize food-contact surfaces; use sanitary water sources	Ensure foodhandlers practice good personal hygiene; obtain shellfish from approved sources; use sanitary, chlorinated water	Ensure foodhandlers practice good personal hygiene; prevent cross-contamination from hands; thoroughly cook food to required minimum internal temperatures; use sanitary, chlorinated water

Exhibit 2e

Major Foodborne Illnesses Caused by Parasites			
Foodborne Illness	**Trichinosis**	**Anisakiasis**	**Giardiasis**
Parasite	*Trichinella spiralis*	*Anisakis simplex*	*Giardia duodenalis* (also called *G. lamblia*)
Symptoms	Nausea, vomiting, diarrhea, fever, and fatigue followed by facial swelling and muscle pain	Tingling or tickling sensation in throat, vomiting or coughing up worms; severe abdominal pain, vomiting, nausea, diarrhea	Intestinal gas, diarrhea, abdominal cramps, nausea, weight loss, fatigue
Source	Domestic pigs; wild game, such as bears and walruses	Marine fish (saltwater species only)	Intestinal tract of humans; contaminated water; inadequately treated water
Food Involved In Outbreaks	Raw and undercooked pork or pork products (particularly sausage), raw and undercooked wild game	Raw, undercooked, or improperly frozen seafood, especially cod, haddock, fluke, Pacific salmon, herring, flounder, halibut monkfish, mackerel and fish used for sashimi and ceviche	Contaminated water and ice, salads and (possibly) other raw vegetables washed with contaminated water
Preventive Measures	Cook pork and game meat to required minimum internal temperatures; wash, rinse, and sanitize equipment, such as sausage grinders and utensils used in the preparation of raw pork and other meats; purchase meat and meat products from approved suppliers; ensure employees practice good personal hygiene	Obtain seafood from approved sources; when serving raw or undercooked seafood, only use sashimi-grade fish that has been properly treated to eliminate parasites; fish intended to be eaten raw should be frozen at −4°F (−20°C) or lower for 7 days in a freezer, or at −31°F (−35°C) or lower for 15 hours in a blast chiller	Use sanitary water supplies; ensure that foodhandlers practice good personal hygiene; wash raw produce carefully

Major Foodborne Illnesses Caused by Parasites *continued*

Foodborne Illness	Toxoplasmosis	Intestinal Cryptosporidiosis	Cyclosporiasis
Parasite	*Toxoplasma gondii*	*Cryptosporidium parvum*	*Cyclospora cayetanensis*
Symptoms	Often, there are no symptoms; when symptoms occur, they include enlarged lymph nodes in head and neck, severe headaches, severe muscle pain, and rash; most commonly affects fetuses	Mild to severe nausea, abdominal cramping, watery diarrhea	Onset of symptoms is sudden; mild to severe nausea, abdominal cramping, mild fever, watery diarrhea
Source	Animal feces (especially felines), mammals	Intestinal tract of humans, cattle, and other domestic animals; drinking water contaminated with run-off from farms or slaughterhouses	Intestinal tract of humans; contaminated water supplies
Food Involved In Outbreaks	Contaminated water; raw or undercooked meat—especially pork, lamb, wild game, and poultry	Water; salads and raw vegetables; milk; unpasteurized apple cider; ready-to-eat food	Water, raw produce, marine fish, raw milk
Preventive Measures	Properly wash hands if they come in contact with soil, raw meat, cat feces, or raw vegetables; avoid raw or undercooked meat (especially lamb, wild game, or poultry); cook meat to the required minimum internal temperature	Ensure that foodhandlers practice good personal hygiene; thoroughly wash produce; use sanitary water sources	Ensure that foodhandlers practice good personal hygiene; thoroughly wash produce; use sanitary water sources

Apply Your Knowledge   **Who Am I?**

Identify the pathogen from
the characteristics given for
each and write its name in the
space provided.

❶ _____

- ▶ I can be found in water contaminated by feces.
- ▶ I am sometimes found in shellfish.
- ▶ I can produce fatigue and jaundice.
- ▶ Obtaining shellfish from an approved source can be a safeguard against me.

❷ _____

- ▶ I can produce intestinal gas.
- ▶ I can come from contaminated water.
- ▶ I can be found in salads.
- ▶ Washing raw produce can prevent me.

❸ _____

- ▶ I can be carried in the intestinal tract of humans.
- ▶ I am sometimes found in sliced melons.
- ▶ I can produce fever and diarrhea in those who ingest me.
- ▶ My growth can be slowed by refrigeration.

For answers, please turn to page 2-24.

FUNGI

Fungi range in size from microscopic, single-celled organisms to very large, multicellular organisms. They are found naturally in air, soil, plants, water, and some food. Mold, yeast, and mushrooms are examples of fungi.

Molds

Molds share some basic characteristics.

▶ They spoil food and sometimes cause illness.

▶ They grow under almost any condition, but grow well in acidic food with low water activity.

▶ Freezing temperatures prevent or reduce the growth of molds, but do not destroy them.

▶ Some molds produce toxins such as aflatoxins.

Although the FDA recommends cutting away any moldy areas in hard cheese—at least one inch (2.5 centimeters) around them—to avoid illnesses caused by mold toxins, throw out all moldy food, unless the mold is a natural part of the food (e.g., cheeses such as Gorgonzola, Bleu, Brie, and Camembert).

While mold cells and spores can be killed by heating them, some toxins are not destroyed by normal cooking methods. Food with molds that are not a natural part of the product should always be discarded.

Yeasts

Some yeasts are known for their ability to spoil food rapidly. Carbon dioxide and alcohol are produced as yeast slowly consumes food. Yeast spoilage may, therefore, produce a smell or taste of alcohol. Yeast may appear as a pink discoloration or slime and may bubble.

Yeasts are similar to molds in that they grow well in acidic food with low water activity, such as jellies, jams, syrup, honey, and fruit juice. (See *Exhibit 2f.*) Food that has been spoiled by yeast should be discarded.

Yeast on Jelly

Food that has been spoiled by yeast should be discarded.

FOODBORNE INFECTION VS. FOODBORNE INTOXICATION

Foodborne diseases are classified as infections, intoxications, or toxin-mediated infections. Each occurs in a different way.

▶ **Foodborne infections** result when a person eats food containing pathogens, which then grow in the intestines and cause illness. Typically, symptoms of a foodborne infection do not appear immediately.

▶ **Foodborne intoxications** result when a person eats food containing toxins that cause illness. The toxin may have been produced by pathogens found on the food or may be the result of a chemical contamination. The toxin might also be a natural part of the plant or animal consumed. Typically, symptoms of a foodborne intoxication appear quickly, within a few hours.

▶ **Foodborne toxin-mediated infections** result when a person eats food containing pathogens, which then produce illness-causing toxins in the intestines.

Apply Your Knowledge   **A Case in Point**

❶ Based on the information given, was the illness caused by bacteria, a virus, a parasite, or fungi?

❷ What is the name of the microorganism most likely to have caused the outbreak?

❸ Is this illness an infection or an intoxication?

For answers, please turn to page 2-24.

A day-care center is serving stir-fried rice for lunch. The rice was cooked to the proper temperature for the proper amount of time at 1:00 P.M. The covered rice was then placed on the countertop and allowed to cool to room temperature. At 6:00 P.M., the cook placed it in the refrigerator. At 9:00 A.M. the following day, the rice was combined with the other ingredients for stir-fried rice and cooked to 165°F (74°C) for at least fifteen seconds. The cook covered the stir-fried rice and left it on the range until she gently reheated it at noon. Within an hour of eating the stir-fried rice, however, several of the children began vomiting, and a few had diarrhea. Samples from some of the children revealed the rice as the probable cause of the outbreak.

SUMMARY

Microbial contaminants are responsible for the majority of foodborne illness. Understanding how microorganisms grow, reproduce, contaminate food, and infect humans is critical to understanding how to prevent the foodborne illnesses they cause.

Of all foodborne microorganisms, bacteria are of greatest concern to the restaurant and foodservice manager. Under favorable conditions, bacteria can reproduce very rapidly. Although vegetative bacteria may be resistant to low—even freezing—temperatures, they can be killed by high temperatures, such as those reached during cooking. Some types of bacteria, however, have the ability to form spores, which protect the bacteria from unfavorable conditions. Since spores are so difficult to destroy, it is important to cook, cool, and reheat food properly.

The acronym FAT TOM—which stands for Food, Acidity, Temperature, Time, Oxygen, and Moisture—is the key to controlling the growth of microorganisms.

Viruses are the smallest of the microbial contaminants. While a virus cannot reproduce in food, once ingested it will cause illness. Practicing good personal hygiene and minimizing bare-hand contact with ready-to-eat food is an important defense against foodborne illness from viruses.

Parasites are organisms that need to live on a host organism to survive. They can live inside many animals humans eat, such as cows, chickens, hogs, and fish. They can be killed by proper cooking and freezing.

Fungi, such as molds and yeasts, are mostly responsible for the spoilage of food. Some molds can produce harmful toxins. Food with molds that are not a natural part of the product should always be discarded. Yeasts are known for their ability to spoil food rapidly. Food spoiled by yeast should also be discarded.

Foodborne diseases are classified as infections, intoxications, or toxin-mediated infections. Foodborne infections result when a person eats food containing pathogens, which then grow in the intestines and cause illness. Typically, symptoms do not appear immediately. Foodborne intoxications result when a person eats food containing toxins produced by pathogens found on the food or which are the result of a chemical contamination. The toxin might also be a natural part of the plant or animal consumed. Typically, symptoms of foodborne intoxication appear quickly, within a few hours. Foodborne toxin-mediated infections result when a person eats food that contains pathogens, which then produce illness-causing toxins in the intestines.

Apply Your Knowledge

Use these questions to test your knowledge of the concepts presented in this section.

Multiple-Choice Study Questions

1. Foodborne microorganisms grow well at temperatures between
   A. 41°F and 135°F (5°C and 57°C).
   B. 32°F and 70°F (0°C and 21°C).
   C. 38°F and 155°F (3°C and 68°C).
   D. 70°F and 165°F (21°C and 74°C).

2. All of the following conditions typically support the growth of microorganisms *except*
   A. moisture.
   B. a protein or carbohydrate food source.
   C. high acidity.
   D. time.

3. A person who has a foodborne infection most likely has eaten a food containing
   A. a ciguatoxin.
   B. histamine.
   C. a plant toxin.
   D. a live pathogen.

4. Which food is most likely to transmit parasites to humans?
   A. Improperly cooked eggs
   B. Improperly frozen sashimi
   C. Improperly refrigerated milk
   D. Unpasteurized apple juice

5. Which of the following is not a basic characteristic of foodborne mold?
   A. It grows well in acidic food with low water activity.
   B. Freezing temperatures prevent or slow its growth, but do not destroy it.
   C. Its cells and spores may be killed by heating, but the toxins it produces may not be destroyed.
   D. It needs a host to survive.

Continued on next page...

6. Which of the following statements regarding foodborne-intoxication is true?
    A. Symptoms of intoxication often appear days after exposure.
    B. Medical treatment for intoxication can be painful.
    C. Foodborne intoxication is more common than foodborne infection.
    D. Symptoms of intoxication appear quickly, within a few hours.

7. Which of the following microorganisms is associated with unpasteurized milk and soft cheeses?
    A. *Vibrio parahaemolyticus*
    B. *Listeria monocytogenes*
    C. *Trichinella spiralis*
    D. *Clostridium botulinum*

8. Which of the following microorganisms is found in cereal crops such as rice?
    A. *Staphylococcus aureus*
    B. Hepatitis A virus
    C. *Campylobacter jejuni*
    D. *Bacillus cereus*

9. A person who has a severe case of *Staphylococcal* Gastroenteritis may experience
    A. changes in blood pressure and pulse rate.
    B. hallucinations.
    C. a tingling or tickling sensation in the throat.
    D. facial swelling.

10. Which of the following practices can help prevent hepatitis A?
    A. Obtaining shellfish from approved sources
    B. Cooking pork to the proper internal temperature
    C. Practicing careful time and temperature control for all thick foods
    D. Avoiding the use of home-canned food

For answers, please turn to page 2-24.

Apply Your Knowledge   Answers

Page	Activity

2-2   Test Your Food Safety Knowledge
1. True
2. False
3. False
4. True
5. True

2-9   What I Need to Grow!
Correct Answer: 2

2-18   Who Am I?
1. Hepatitis A virus
2. *Giardia duodenalis*
3. *Salmonella* spp.

2-20   A Case in Point
❶ Bacteria ❷ *Bacillus cereus* was the microorganism responsible for the outbreak. ❸ Given the rapid onset and the symptoms, the illness was most likely an intoxication.

2-22   Multiple-Choice Study Questions
1. A   6. D
2. C   7. B
3. D   8. D
4. B   9. A
5. D   10. A

Apply Your Knowledge   Notes

Contamination, Food Allergens, and Foodborne Illness

Inside this section:
▶ Types of Foodborne Contamination
▶ The Deliberate Contamination of Food
▶ Food Allergens

After completing this section, you should be able to:

▶ Identify biological, chemical, and physical contaminants.
▶ Identify methods to prevent biological, chemical, and physical contamination.

▶ Identify the eight most common allergens, associated symptoms, and methods of prevention.

Apply Your Knowledge	Test Your Food Safety Knowledge

Check to see how much you know about the concepts in this section. Use the page references provided to explore the topic in each question.

❶ **True or False:** Fish that has been properly cooked will be safe to eat. *(See page 3-4.)*

❷ **True or False:** Cooking can destroy the toxins in toxic wild mushrooms. *(See page 3-5.)*

❸ **True or False:** Copper utensils and equipment can cause an illness when used to prepare acidic food. *(See page 3-7.)*

❹ **True or False:** Cleaning products may be stored with packages of food. *(See page 3-8.)*

❺ **True or False:** A person who is allergic to food may experience tightening in the throat. *(See page 3-12.)*

For answers, please turn to page 3-18.

CONCEPTS

▶ **Biological contaminant:** Microbial contaminant that may cause foodborne illness. These contaminants include bacteria, viruses, parasites, fungi, and biological toxins.

▶ **Chemical contaminant:** Chemical substance that can cause a foodborne illness. Food can become contaminated by a variety of chemical substances normally found in restaurant and foodservice establishments, including toxic metals, pesticides, cleaning products, sanitizers, and lubricants.

▶ **Physical contaminant:** Foreign object that is accidentally introduced into food, or a naturally occurring object, such as a bone in a filet, that poses a physical hazard. Common physical contaminants include metal shavings from cans, staples from cartons, glass from broken light bulbs, blades from plastic or rubber scrapers, fingernails, hair, bandages, dirt, and bones.

▶ **Biological toxins:** Toxins (poisons) produced by pathogens, plants, or animals. They may also occur in animals as a result of their diet.

▶ **Ciguatera poisoning:** Illness that occurs when a person eats fish that has consumed the ciguatera toxin. This toxin occurs in certain predatory tropical reef fish, such as amberjack, barracuda, grouper, and snapper.

▶ **Scombroid poisoning:** Illness that occurs when a person eats a scombroid fish that has been time-temperature abused. Scombroid fish include tuna, mackerel, bluefish, skipjack, and bonito.

▶ **Food security:** The prevention or elimination of the deliberate contamination of food.

INTRODUCTION

Food is considered contaminated when it contains hazardous substances. These substances may be biological, chemical, or physical. The most common food contaminants are biological contaminants that belong to the microworld—bacteria, parasites, viruses, and fungi. Most foodborne illnesses result from these contaminants, but biological and chemical toxins are also responsible for many foodborne illnesses. While biological and chemical contamination pose a significant threat to food, the danger from physical hazards should also be recognized.

TYPES OF FOODBORNE CONTAMINATION

A thorough understanding of the causes and prevention of various types of contamination can help you keep food safe.

Biological Contamination

As you learned in Section 2, a foodborne intoxication occurs when a person eats food containing toxins. The toxin may have been produced by pathogens found on the food or may be the result of a chemical contamination. The toxin could also come from a plant or animal that was eaten. Toxins in seafood, plants, and mushrooms are responsible for many cases of foodborne illness in the U.S. each year. Most of these biological toxins occur naturally and are not caused by the presence of microorganisms. Some occur in animals as a result of their diet.

Seafood Toxins

▶ The ciguatera toxin occurs in certain predatory tropical reef fish, such as amberjack, barracuda, grouper, and snapper. Ciguatera accumulates in the tissue of these large, predatory fish after they eat smaller fish that have fed upon certain species of toxic algae. Some important points about ciguatera include:

▶ Eating fish containing this toxin may result in ciguatera poisoning.

▶ Symptoms of ciguatera poisoning include vomiting, severe itching, nausea, dizziness, hot and cold flashes, temporary blindness, and sometimes hallucinations.

▶ The ciguatera toxin cannot be smelled or tasted and it is not destroyed by cooking. Therefore, it is very important to purchase predatory tropical reef fish only from approved suppliers.

▶ Shellfish may contain toxins that occur because of the algae upon which they feed. Illness caused by shellfish toxins varies and is specific to the type of toxin consumed.

▶ Since cooking might not destroy shellfish toxins, it is important to purchase shellfish from approved suppliers who can certify the shellfish have been harvested from safe waters.

▶ Scombroid poisoning is one of the most common forms of illness caused by fish toxins in the U.S. Some important points about scombroid poisoning include:

▶ It occurs when scombroid species of fish—tuna, mackerel, bluefish, skipjack, and bonito—are time-temperature abused. Under these conditions, bacteria associated with these fish produce the toxin, histamine.

▶ Symptoms of the illness include flushing and sweating, a burning, peppery taste in the mouth, dizziness, nausea, and headache. Sometimes a facial rash, hives, edema, diarrhea, and abdominal cramps will follow.

▶ Histamine is not destroyed by cooking or freezing.

▶ Since time-temperature abuse during the harvesting process may cause scombroid fish to become unsafe, it is important to purchase these fish from reputable suppliers who practice strict time-temperature control.

To guard against seafood-specific foodborne illness, always purchase from reputable suppliers. Check the temperature of fish upon delivery, making sure it has been received at 41°F (5°C) or lower. Refuse fish that has been thawed and refrozen.

Mushroom Toxins

▶ Foodborne-illness outbreaks associated with mushrooms are almost always caused by the consumption of wild mushrooms collected by amateur mushroom hunters. Most cases occur when toxic mushroom species are confused with edible species.

▶ The symptoms of intoxication vary depending upon the species consumed.

▶ Cooking or freezing will not destroy toxins produced by toxic wild mushrooms.

Foodservice establishments should not use mushrooms picked in the wild or products made with them unless the mushrooms have been purchased from approved suppliers. Establishments that serve mushrooms picked in the wild should have written buyer specifications that

▶ identify the mushroom's common name, and the Latin binomial and its author.

▶ ensure that the mushroom was identified in its fresh state.

▶ indicate the name of the person who identified the mushroom and include a statement regarding his/her qualifications.

See *Exhibit 3a* on the next page for a summary of common biological toxins.

Exhibit 3a

Biological Toxins (Biological Contaminants)			
Biological Toxin	**Source of Contamination**	**Associated Food**	**Preventive Measures**
Seafood Toxins — **Ciguatera Toxin**	Fish that have eaten algae containing the toxin	Predatory tropical reef fish, such as amberjack, barracuda, grouper, and snapper	Cooking does not destroy these toxins; purchase predatory tropical reef fish only from approved suppliers
Scombroid Toxin (histamine)	Histamine produced by bacteria in some fish when they are time-temperature abused	Primarily occurs in tuna, bluefish, mackerel, skipjack, roundfish, and bonito; other fish, such as mahi-mahi, marlin, and sardines, have also been implicated in histamine poisoning	Cooking does not destroy histamine; because time-temperature abuse during the harvesting process may cause the fish to become unsafe, it is important to purchase from reputable suppliers
Shellfish Toxins	Shellfish that have eaten a type of algae containing the toxin	Shellfish, especially mollusks, such as mussels, clams, cockles, and scallops	Cooking may not destroy these toxins; purchase these shellfish from approved suppliers who can certify they are harvested from safe waters
Systemic Fish Toxins	Toxins that are a natural part of some fish	Pufferfish, moray eels, and freshwater minnows	Cooking may not destroy systemic fish toxins; pufferfish should be handled and prepared by properly trained chefs
Plant Toxins	Toxins that are a natural part of some plants	Fava beans, rhubarb leaves, jimsonweed, water hemlock, and apricot kernels; honey from bees that have gathered nectar from mountain laurel; milk from cows that have eaten snakeroot, jimsonweed, or other toxic plants	Cooking may not destroy these toxins; avoid these plant species and products prepared with them
Fungal Toxins	Toxins that are a natural part of some varieties of fungi	Poisonous varieties of mushrooms and other fungi	Cooking does not destroy these toxins; do not use mushrooms picked in the wild or products made with them unless the mushrooms have been purchased from approved suppliers

Apply Your Knowledge

Who Am I?

Identify the toxin in each situation and write its name in the space provided.

① _____

▶ I accumulate in the tissue of large predatory fish.

▶ I can produce severe itching, temporary blindness, and hallucinations.

▶ I am not destroyed by cooking.

② _____

▶ I can occur when tuna and mackeral are time-temperature abused.

▶ I can produce a burning, peppery taste in the mouth.

▶ Purchasing fish from reputable suppliers can help prevent me.

For answers, please turn to page 3-18.

Exhibit 3b

Toxic Metals

Acidic food prepared in equipment made from toxic metals can cause illness.

Chemical Contamination

Chemical contaminants are responsible for many cases of foodborne illness. Contamination can come from a variety of substances normally found in restaurants and foodservice establishments. These include toxic metals, pesticides, cleaning products, sanitizers, and lubricants.

Toxic Metals

▶ Utensils and equipment that contain toxic metals—such as lead, copper, brass, zinc, antimony, and cadmium—can cause toxic-metal poisoning. If acidic food is stored in or prepared with this type of equipment, it can leach these metals from the item and become contaminated. (See *Exhibit 3b.*)

▶ Only food-grade utensils and equipment should be used to prepare and store food.

Chemicals and Pesticides

▶ Chemicals such as cleaning products, polishes, lubricants, and sanitizers can contaminate food if they are improperly used or stored.

▶ Follow the directions supplied by the manufacturer when using chemicals.

▶ Exercise caution when using chemicals during operating hours to prevent contamination of food and food-preparation areas.

▶ Store chemicals away from food, utensils, and equipment used for food. Keep them in a locked storage area in their original container.

▶ If chemicals must be transferred to smaller containers or spray bottles, label each container appropriately.

▶ Pesticides are often used in kitchens and in food-preparation and storage areas to control pests, such as rodents and insects. All food should be wrapped or stored prior to application of the pesticide. If pesticides are stored in the establishment, exercise the same care as with other chemicals used there.

See *Exhibit 3d* for a summary of common chemical contaminants.

Physical Contamination

▶ Physical contamination results when foreign objects are accidentally introduced into food, or when naturally occurring objects, such as bones in fillets, pose a physical hazard.

▶ Common physical contaminants may include metal shavings from cans, staples from cartons, glass from broken light bulbs, blades from plastic or rubber scrapers, fingernails, hair, bandages, dirt, and bones. (See *Exhibit 3c*.)

▶ Closely inspect the food you receive and take steps to ensure it will not be physically contaminated during the flow of food in your operation.

Exhibit 3c

Physical Contamination

Metal shavings in an opened can might contaminate the food inside.

Exhibit 3d

Chemical Contaminants

Chemical Toxin	Source of Contamination	Associated Food	Preventive Measures
Toxic Metals	Utensils and equipment containing potentially toxic metals, such as lead, copper, brass, zinc, antimony, and cadmium	▶ Any food, but especially high-acid food, such as sauerkraut, tomatoes, and citrus products; the acidity of this food can cause metal ions to leach into its liquids ▶ Carbonated beverages; the carbonated water used to make the beverage might leach copper ions from copper, water-supply lines	▶ Use only food-grade storage containers ▶ Use metal and plastic containers only for their intended use ▶ Use only food-grade brushes on food; do not use paintbrushes or wire brushes ▶ Do not use enamelware, which may chip and expose the underlying metal ▶ Do not use equipment or utensils made of materials that contain lead (such as pewter) for food preparation ▶ Do not use zinc-coated (galvanized) equipment or utensils for food preparation ▶ Use a backflow-prevention device to prevent carbonated water in soft drink-dispensing systems from flowing back into the copper water-supply lines
Chemicals	Cleaning products, polishes, lubricants, and sanitizers		▶ Follow manufacturers' directions for storage and use; use only recommended amounts ▶ Store away from food, utensils, and equipment used for food ▶ Store in a dry cabinet in original, labeled containers, apart from other chemicals that might react with them ▶ Tools used for dispensing chemicals should never be used on food ▶ If chemicals must be transferred to smaller containers or spray bottles, label each container appropriately ▶ Use only food-grade lubricants or oils on kitchen equipment or utensils
Pesticides	Used in kitchens and food-preparation and storage areas to control pests, such as rodents and insects		▶ Pesticides should only be applied by a licensed professional; wrap or store all food before pesticides are applied

THE DELIBERATE CONTAMINATION OF FOOD

While the principles of food safety help an establishment address the accidental contamination of food, managers must also be aware of how to prevent or eliminate the deliberate contamination of food, known as food security. In addition to biological, chemical, and physical contaminants, nuclear and radioactive contaminants are also a concern.

Threats to food security in the restaurant and foodservice industry might occur at any level in the food-supply chain and are the result of criminal activity. These attacks are usually focused on a specific food item, process, company, or business. Those who would knowingly contaminate a food product include, but are not limited to, organized terrorist or activist groups, individuals posing as customers, current or former employees, vendors, and competitors.

The key to protecting food is to make it as difficult as possible for even a single tampering to occur. An effective food security program will consider all of the points where food is vulnerable to intentional contamination. Potential threats can come from the following three areas:

▶ Human elements

▶ Interior elements

▶ Exterior elements

Managers must ensure that all employees are aware of their roles in keeping food secure in the operation by developing procedures and training that address each potential threat. *Exhibit 3e* lists elements to consider when determining how to handle food security in your establishment.

Exhibit 3e

Addressing Food Security Threats in Your Operation*

Human Elements

► Learn about your applicants—ask for references, verify references, and check identification.

► Train employees in food security and establish food security awareness with all employees.

► Train your employees to report any suspicious activity.

► Establish a system to identify employees on duty. Only on-duty employees should be allowed in work areas.

► Establish rules for the opening of the back door—determine those employees that are authorized to open these doors and under what circumstances.

► Control customer and non-employee access to food-production areas.

► Allow employees to bring only essential items to work. These items might include uniforms, name badges, and anything else necessary for job functions.

► Consider a two-employee rule during food preparation—no single employee should be allowed in a production area by him/herself. Having more than one employee around is a built-in check system.

► Supervise and survey production areas on a regular basis. This can be accomplished via video cameras, windows, other employees, or management.

Interior Elements

► Limit access to doors, windows, roofs, and food-storage areas. Good lighting is also important.

► Keep food display, storage, and kitchen entrances and exits controlled or under surveillance.

► Eliminate hiding places in all areas of the operation. Make sure there are no places for an intruder to hide until after working hours.

► Identify and inspect all incoming food items, and never accept suspect food. Have a specific food inspection program in place, and make sure that employees are familiar with it.

► Restrict traffic in food-prep and food-storage areas.

► Monitor all customer self-service and displayed food items, such as salad bars, condiments, and exposed tableware.

Exterior Elements

► Ensure that the building's exterior is well lit. There should be no areas where an intruder could remain unseen.

► Control access to ventilation system to prevent tampering.

► Identify all company food suppliers and consider using tamper-evident packages. Check the identification of the delivery person, the scheduled times of delivery, and document those deliveries.

► Tell your suppliers that food security is a priority, and ask what steps they are taking to ensure their products are secure.

► Verify and pre-approve all service personnel and providers.

► Prevent unmonitored access to facilities by non-employees after normal business hours.

*For more information on food security, visit www.nraef.org/foodsecurity.

FOOD ALLERGENS

Six to seven million Americans have food allergies. A food allergy is the body's negative reaction to a particular food protein. Depending on the person, allergic reactions may occur immediately after the food is eaten or several hours later. The reaction could include some or all of the following symptoms:

▶ Itching in and around the mouth, face, or scalp

▶ Tightening in the throat

▶ Wheezing or shortness of breath

▶ Hives

▶ Swelling of the face, eyes, hands, or feet

▶ Gastrointestinal symptoms, including abdominal cramps, vomiting, or diarrhea

▶ Loss of consciousness

▶ Death

Employees should be aware of the most common food allergens, including milk and dairy products, eggs and egg products, fish, shellfish, wheat, soy and soy products, peanuts, and tree nuts.

Your employees should be able to inform customers of menu items that contain these potential allergens. Designate one person per shift to answer customers' questions regarding menu items. To help customers with allergies enjoy a safe meal at your establishment, keep the following points in mind:

▶ Be able to fully describe each of your menu items when asked. Tell customers how the item is prepared and identify any "secret" ingredients used.

▶ If you do not know if an item is free of an allergen, tell the customer. Urge the customer to order something else.

▶ When preparing food for a customer with allergies, ensure that the food makes no contact with the ingredient to which the customer is allergic. Make sure all cookware, utensils, and tableware are allergen-free to prevent food contamination.

▶ Serve menu items as simply as possible to customers with allergies. Sauces and garnishes are often the source of allergic reactions. Serve these items on the side.

Apply Your Knowledge

Place an **X** next to the food items that are common allergens.

Spot the Allergen

__ Eggs   __ Bean sprouts

__ Peanuts   __ Shellfish

__ Beef   __ Milk

For answers, please turn to page 3-18.

Apply Your Knowledge

❶ Explain why the mahi-mahi steaks were implicated in an outbreak of scombroid poisoning.

For answers, please turn to page 3-18.

A Case in Point

Roberto receives a shipment of frozen mahi-mahi steaks. The steaks are frozen solid at the time of delivery and the packages are sealed and contain a lot of ice crystals, indicating they have been time-temperature abused. Roberto accepts the mahi-mahi steaks and thaws them in the refrigerator at a temperature of 38°F (3°C). The thawed fish steaks are then held at this temperature during the evening shift and are cooked to order. The cooks follow all appropriate guidelines for preparing, cooking, and serving the fish, monitoring time and temperature throughout the process. Unfortunately, these fish steaks are implicated in an outbreak of scombroid poisoning.

SUMMARY

Biological and chemical toxins are responsible for many foodborne-illness outbreaks. Most occur naturally and are not caused by the presence of microorganisms. Some occur in the animal as a result of its diet. Since toxins are not living organisms, cooking or freezing typically will not destroy them. Most measures taken to prevent foodborne intoxication center on proper purchasing and receiving.

Purchase seafood from reputable suppliers who maintain strict time-temperature controls and can certify the seafood has been harvested from safe waters. Do not use mushrooms picked in the wild or products made with them unless the mushrooms have been purchased from approved suppliers.

Use only food-grade utensils and equipment to prepare and store food. Cleaning products, polishes, lubricants, and sanitizers should be used as directed. Exercise caution when using these chemicals during operating hours and store them properly. If used, pesticides should be applied by a licensed professional. Physical contamination can occur when physical objects are accidentally introduced into food or when naturally occurring objects, such as the bones in fish, pose a physical hazard. Closely inspect the food you receive and take steps to ensure food will not become physically contaminated during its flow through your operation.

Food security addresses the prevention or elimination of the deliberate contamination of food. Contamination can occur in biological, chemical, physical, nuclear, or radioactive form. The key to protecting food is to make it as difficult as possible for even a single tampering to occur. Managers must develop and maintain a food security program that focuses on the potential threats posed by the interior, exterior, and human elements of their establishment.

Many people have food allergies. Employees should be aware of the most common food allergens, which include milk and dairy products, eggs and egg products, fish, shellfish, wheat, soy and soy products, peanuts, and tree nuts. You should be able to inform customers of these and other potential food allergens that may be included in food served at your establishment.

Apply Your Knowledge

Draw Your Own Conclusion

Read the following scenario. When you are finished, your job is to help Paul figure out what foodborne illness made his customers sick and what caused the illness. To narrow down your choice, you can ask Paul questions about the incident. Your instructor will play the role of Paul. Try to crack the case while asking the least number of questions. Good luck!

For answers, please see your instructor.

Scenario: Paul is the owner of Tropical Reef Hideaway, a restaurant in a small coastal town. Today, he received several foodborne-illness complaints from customers who recently dined at his restaurant. This surprised Paul, since the establishment follows good food safety practices. The customers all dined in his establishment on the same day. Paul's investigation of the incident has revealed the following facts:

▶ The doors of the restaurant were left open several times during the day, which let in a large number of flies.

▶ The customers all ate raw oysters on the half-shell as an appetizer.

▶ Fresh fish was purchased the day of the incident.

▶ The customers who became ill complained of nausea, dizziness, and severe itching.

▶ The drink of the day was a Mai Tai (a fresh fruit juice-based alcoholic beverage), made in large quantities. All of the customers ordered a Mai Tai.

▶ All of the fish served that evening was cooked to the required minimum internal temperature.

Apply Your Knowledge

Use these questions to test your knowledge of the concepts presented in this section.

Multiple-Choice Study Questions

1. You have ordered frozen tuna steaks for your restaurant. When the delivery arrives, you notice there is excessive frost and ice in the package, which indicates they have been time-temperature abused. You refuse the delivery. Why?
   A. You suspect the steaks may have been contaminated with a cleaning compound.
   B. You suspect the steaks may contain ciguatera toxins.
   C. You suspect the steaks may cause scombroid poisoning if you serve them.
   D. You believe the supply company may have treated the steaks with an unauthorized preservative.

2. A customer becomes ill after eating grouper. It is discovered that the shipment of grouper contained the ciguatera toxin. This is an example of
   A. chemical contamination.
   B. physical contamination.
   C. biological contamination.

3. Which of the following is not a common food allergen?
   A. Eggs
   B. Dairy products
   C. Peanuts
   D. Beef

4. You find a piece of glass at the bottom of your ice storage bin. This is an example of
   A. chemical contamination.
   B. physical contamination.
   C. biological contamination.

5. Which of the following fish is associated with ciguatera?
   A. Mahi-mahi
   B. Pufferfish
   C. Snapper
   D. Fresh-water minnow

Continued on next page...

6. An establishment should do all of the following to guard against a seafood-specific foodborne illness *except*

    A. thaw fish at temperatures higher than 41°F (5°C).

    B. purchase seafood from suppliers who practice time-temperature control.

    C. refuse fish that has been thawed and refrozen.

    D. accept fish that has been received at 41°F (5°C) or lower.

7. An establishment should do all of the following to prevent contamination *except*

    A. store food away from chemicals.

    B. keep high-acid food separate from other types of food.

    C. purchase food products from approved suppliers.

    D. use food-grade storage containers.

8. Which of the following is an example of a physical contaminant?

    A. A virus present in shellfish

    B. Dirt on a head of lettuce

    C. Sanitizer residue left on a cutting board

    D. Parasites present in a raw fish fillet

9. All of the following can lead to the chemical contamination of food *except*

    A. cooking tomato sauce in a copper pot.

    B. storing orange juice in a pewter pitcher.

    C. using a backflow-prevention device on a carbonated beverage dispenser.

    D. serving fruit punch in a galvanized tub.

10. Which of the following statements is true about fish containing ciguatera or scombroid (histamine) toxins?

    A. Freezing will destroy these toxins.

    B. Cooking will not destroy these toxins.

    C. You can see and smell these toxins.

    D. Cooking will destroy these toxins.

For answers, please turn to page 3-18.

Apply Your Knowledge   Answers

Page	Activity

3-2   Test Your Food Safety Knowledge
1. False   2. False   3. True   4. False   5. True

3-7   Who Am I?
1. Ciguatera
2. Histamine

3-13   Spot the Allergen

Eggs, Peanuts, Shellfish, Milk

3-13   A Case in Point

❶ When the mahi-mahi (a scombroid species of fish) was time-temperature abused, the bacteria associated with the fish produced the toxin histamine. Since cooking does not destroy this toxin, the consumption of the fish resulted in a scombroid poisoning (intoxication).

3-16   Multiple-Choice Study Questions
1. C   6. A
2. C   7. B
3. D   8. B
4. B   9. C
5. C   10. B

Apply Your Knowledge   Notes

▶ **Single-use gloves:** Disposable gloves designed for one-time use to provide a barrier between hands and the food they come in contact with. Gloves should never be used in place of handwashing. Foodhandlers should wash hands before putting on gloves and when changing to a fresh pair.

▶ **Jaundice:** Yellowing of the skin and eyes that could indicate a person is ill with hepatitis A.

INTRODUCTION

At every step in the flow of food through the operation—from purchasing and receiving through final service—foodhandlers can contaminate food and cause customers to become ill. Good personal hygiene is a critical protective measure against foodborne illness, and customers expect it.

You can minimize the risk of foodborne illness by establishing a personal hygiene program that spells out your specific hygiene policies, provides your employees with training on those policies, and enforces established policies. When employees have the proper knowledge, skills, and attitudes toward personal hygiene, you are one step closer to operating a safe food system.

HOW FOODHANDLERS CAN CONTAMINATE FOOD

Earlier, you learned that foodhandlers can cause illness when they transfer microorganisms to food they touch. Many times these microorganisms come from the foodhandlers themselves. Foodhandlers can contaminate food when they

▶ have a foodborne illness.

▶ show symptoms of gastrointestinal illness (an illness relating to the stomach or intestine).

▶ have infected lesions (wounds or injuries).

▶ live with, or are exposed to, a person who is ill.

▶ touch anything that may contaminate their hands.

Even an apparently healthy person may be hosting foodborne pathogens. With some illnesses, such as hepatitis A, a person is at the most infectious stage of the disease for several

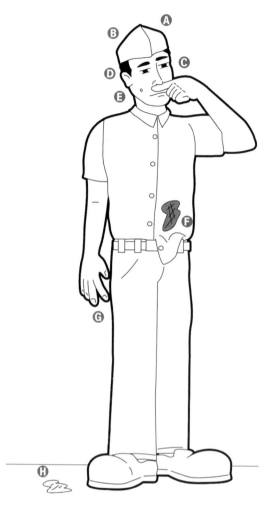

weeks before symptoms appear. With other illnesses, the pathogens may remain in a person's system for months after all signs of infection have ceased. Some people are called carriers because they might carry pathogens and infect others, yet never become ill themselves.

Simple acts or personal behaviors can contaminate food. Since it is so easy to contaminate food, foodhandlers must pay close attention to what they do with their hands and maintain good personal hygiene. Actions to avoid include:

- **Ⓐ** Scratching the scalp
- **Ⓑ** Running fingers through hair
- **Ⓒ** Wiping or touching the nose
- **Ⓓ** Rubbing an ear
- **Ⓔ** Touching a pimple or an open sore
- **Ⓕ** Wearing a dirty uniform
- **Ⓖ** Coughing or sneezing into the hand
- **Ⓗ** Spitting in the establishment

DISEASES NOT TRANSMITTED THROUGH FOOD

In recent years, the public has expressed growing concern over communicable diseases spread through intimate contact or by direct exchange of bodily fluids. Diseases such as AIDS (Acquired Immune Deficiency Syndrome), hepatitis B and C, and tuberculosis are not spread through food.

COMPONENTS OF A GOOD PERSONAL HYGIENE PROGRAM

Good personal hygiene is key to the prevention of foodborne illness and includes:

Ⓐ Maintaining personal cleanliness

▶ Proper bathing

▶ Hair washing

Ⓑ Wearing proper work attire

▶ Clean hat or hair restraint

▶ Clean clothing

▶ Appropriate shoes

▶ Removing jewelry

Ⓒ Following hygienic hand practices

▶ Handwashing

▶ Hand maintenance

▶ Proper glove use

Employees must also avoid unsanitary habits and actions, maintain good health, and report any illness or injury.

Hygienic Hand Practices

Handwashing

While handwashing may appear fundamental, many foodhandlers fail to wash their hands properly and as often as needed. As a manager, it is your responsibility to train your foodhandlers and then monitor them. Never take this simple action for granted. To ensure proper handwashing in your establishment, train your foodhandlers to follow the steps illustrated in *Exhibit 4a* on the next page.

Hand sanitizers (liquids used to lower the number of microorganisms on the skin surface) or hand dips may be used after washing, but should never be used in place of proper handwashing. If hand sanitizers are used, foodhandlers should never touch food or food-preparation equipment until the hand

Proper Handwashing Procedure

❶ Wet your hands with running water as hot as you can comfortably stand (at least 100°F/38°C).

❷ Apply soap.

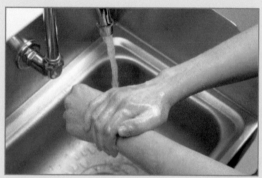

❸ Vigorously scrub hands and arms for at least twenty seconds.

❹ Clean under fingernails and between fingers.

❺ Rinse thoroughly under running water.

❻ Dry hands and arms with a single-use paper towel or warm-air hand dryer.

sanitizer has dried. Establishments must only use hand sanitizers that have been approved by the FDA.

Foodhandlers must wash their hands before they start work and after the following activities:

▶ Using the restroom

▶ Handling raw food (before *and* after)

▶ Touching the hair, face, or body

▶ Sneezing, coughing, or using a handkerchief or tissue

▶ Smoking, eating, drinking, or chewing gum or tobacco

▶ Handling chemicals that might affect the safety of food

▶ Taking out garbage or trash

▶ Clearing tables or busing dirty dishes

▶ Touching clothing or aprons

▶ Touching anything else that may contaminate hands, such as unsanitized equipment, work surfaces, or washcloths

---

Apply Your Knowledge | **Put the Steps in Order**

Put the handwashing steps in order by placing the number of the step in the space provided.

____ Ⓐ Vigorously scrub hands and arms for at least twenty seconds.

____ Ⓑ Wet hands with running water as hot as you can comfortably stand (at least 100°F/38°C).

____ Ⓒ Rinse thoroughly under running water.

____ Ⓓ Clean under fingernails and between fingers.

____ Ⓔ Apply soap.

____ Ⓕ Dry hands and arms with a single-use paper towel or warm-air hand dryer.

For answers, please turn to page 4-23.

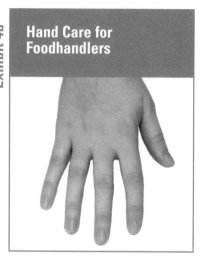

Keep fingernails short and clean.

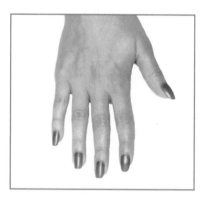

Do not wear false nails or nail polish.

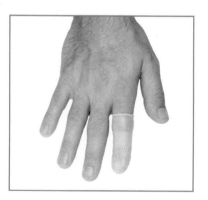

Bandage cuts and cover bandages.

Bare-Hand Contact with Ready-To-Eat Food

Proper handwashing minimizes the risk of contamination associated with bare-hand contact with ready-to-eat food. For those jurisdictions that allow bare-hand contact with this food, establishments must have a verifiable written policy on handwashing procedures. **Check with your regulatory agency for requirements in your jurisdiction.**

Hand Maintenance

In addition to proper washing, hands need other regular care to ensure that they will not transfer microorganisms to food. To keep food safe, make sure foodhandlers follow these guidelines (see *Exhibit 4b*):

▶ **Keep fingernails short and clean.** Long fingernails, may be difficult to keep clean.

▶ **Do not wear false fingernails.** False and acrylic nails should not be worn while handling food since they can be difficult to keep clean and can break off into food. Some jurisdictions allow single-use gloves to be worn over false nails. **Check your local requirements.**

▶ **Do not wear nail polish.** It can disguise dirt under nails and may flake off into food. Some jurisdictions allow single-use gloves to be worn over polished nails. **Check your local requirements.**

▶ **Cover all hand cuts and sores with clean bandages.** If hands are bandaged, clean gloves or finger cots, a protective covering, should be worn at all times to protect the bandage and to prevent it from falling off into food. You may need to move the foodhandler to another job, where he or she will not handle food or touch food-contact surfaces, until the injury heals.

Glove Use

Gloves can help keep food safe by creating a barrier between hands and food. When purchasing gloves for handling food, managers should:

▶ **Buy the right glove for the task.** Long gloves, for example, should be used for hand-mixing salads. Colored gloves can also be used to help prevent cross-contamination.

▶ **Provide a variety of glove sizes.** Gloves that are too big will not stay on the hand and those that are too small will tear or rip easily.

▶ **Consider latex alternatives for employees who are sensitive to the material.**

▶ **Focus on safety, durability, and cleanliness.** Make sure you purchase gloves specifically formulated for food contact, which include gloves bearing the NSF certification mark.

Gloves must never be used in place of handwashing. Hands must be washed before putting on gloves and when changing to a fresh pair. Gloves used to handle food are for single use only and should never be washed and re-used. They should be removed by grasping them at the cuff and peeling them off inside out over the fingers while avoiding contact with the palm and fingers. Foodhandlers should change their gloves

▶ as soon as they become soiled or torn.

▶ before beginning a different task.

▶ at least every four hours during continual use, and more often when necessary.

▶ after handling raw meat and before handling cooked or ready-to-eat food.

Maintaining Personal Cleanliness

In addition to following proper hand hygiene practices, foodhandlers must maintain personal cleanliness. They should bathe or shower before work. Foodhandlers must also keep their hair clean, since oily, dirty hair can harbor pathogens.

Proper Work Attire

A foodhandler's attire plays an important role in the prevention of foodborne illness. Dirty clothes may harbor pathogens and give customers a bad impression of your establishment. Therefore, managers should make sure that foodhandlers observe strict dress standards.

Foodhandlers should:

Ⓐ **Wear a clean hat or other hair restraint.** A hair restraint will keep hair away from food and keep the foodhandler from touching it. Foodhandlers with facial hair should also wear beard restraints.

Ⓑ **Wear clean clothing daily.** If possible, foodhandlers should put on their work clothes at the establishment.

Ⓒ **Remove aprons when leaving food-preparation areas.** For example, aprons should be removed and properly stored prior to taking out garbage or using the restroom.

Ⓓ **Remove jewelry prior to preparing or serving food, or while working around food-preparation areas.** Jewelry can harbor microorganisms, can tempt foodhandlers to touch it, and may pose a safety hazard around equipment. Remove rings (except for a plain band), bracelets (including medical information jewelry), watches, earrings, necklaces, and facial jewelry (such as nose rings, etc.).

Ⓔ **Wear appropriate shoes.** Wear clean, closed-toe shoes with a sensible, nonslip sole.

Check with your local regulatory agency regarding requirements. These requirements should be reflected in written policies, which should be consistently monitored and enforced. All potential employees should be made aware of these policies prior to employment.

Policies Regarding Eating, Drinking, Chewing Gum, and Tobacco

Small droplets of saliva can contain thousands of disease-causing microorganisms. In the process of eating, drinking, chewing gum, or smoking, saliva can be transferred to the foodhandler's hands or directly to food being handled.

Foodhandlers must not:

▶ Smoke, chew gum or tobacco, eat, or drink

When:

▶ Preparing or serving food

▶ In food-preparation areas

▶ In areas used to clean utensils and equipment

Some jurisdictions allow employees to drink from a covered container with a straw while in these areas. **Check with your local regulatory agency.** Foodhandlers should eat, drink, chew gum, or use tobacco products only in designated areas, such as an employee break room. They should never be allowed to spit in the establishment.

If food must be tasted during preparation, it must be placed in a separate dish and tasted with a clean utensil. The dish and utensil should then be removed from the food-preparation area for cleaning and sanitizing.

Policies for Reporting Illness and Injury

Foodhandlers must report health problems to the manager of the establishment before working with food. If they become ill while working, they must immediately report their condition and if food or equipment could become contaminated, the foodhandler must stop working and see a doctor. There are several instances when a foodhandler must either be restricted from working with or around food or excluded from working within the establishment. (See *Exhibit 4c* on the next page.)

Handling Employee Illnesses

If	Then
The foodhandler has one of the following symptoms: ▶ Fever ▶ Diarrhea ▶ Vomiting ▶ Sore throat with fever ▶ Jaundice (a yellowing of the skin and eyes)	Restrict them from working with or around food. Exclude them from the establishment if you primarily serve a high-risk population.
The foodhandler has been diagnosed with a foodborne illness.	Exclude them from the establishment and notify the local regulatory agency. Managers must report employee illnesses resulting from the following pathogens to the local health department: ▶ *Salmonella typhi* ▶ *Shigella* spp. ▶ Shiga toxin-producing *E. coli* ▶ Hepatitis A virus The manager must work with the local regulatory agency to determine when the foodhandler can safely return to work.

If the foodhandler must refrigerate personal medication while working, it must be stored inside a covered, leak-proof container that is clearly labeled.

Any cuts, burns, boils, sores, skin infections, or infected wounds should be covered with a bandage when the foodhandler is working with or around food or food-contact surfaces. Bandages should be clean, dry, and must prevent leakage from the wound. As previously mentioned, waterproof disposable gloves or finger cots should be worn over bandages on hands. Foodhandlers wearing bandages may need to be temporarily reassigned to duties not involving contact with food or food-contact surfaces.

Apply Your Knowledge

Write an **E** in the space next to the statement if the foodhandler must be excluded from the establishment or an **R** if he or she should be restricted from working with or around food.

Exclusion or Restriction?

___ ❶ Bill, a line cook at a family restaurant, has a sore throat with a fever.

___ ❷ Joe, a hospital prep cook, has diarrhea.

___ ❸ Mary, a sous chef, has been diagnosed with hepatitis A.

For answers, please turn to page 4-23.

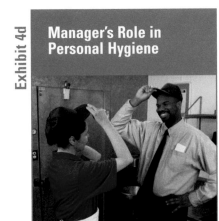

Exhibit 4d

Manager's Role in Personal Hygiene

Managers must model proper behavior for foodhandlers at all times.

MANAGEMENT'S ROLE IN A PERSONAL HYGIENE PROGRAM

Management plays a critical role in the effectiveness of a personal hygiene program. (See *Exhibit 4d.*) Your responsibilities include:

▶ Establishing proper personal hygiene policies

▶ Training foodhandlers on personal hygiene policies

▶ Modeling proper behavior for foodhandlers at all times

▶ Supervising sanitary practices continuously, and retraining foodhandlers as necessary

▶ Revising policies when laws and regulations change, as well as when changes are recognized in the science of food safety, and retraining foodhandlers as necessary

Apply Your Knowledge

What's Wrong with This Picture?

There are at least thirteen unsafe foodhandling practices in this picture. Identify them in the space provided.

For answers, please turn to page 4-23.

❶ _____

❷ _____

❸ _____

❹ _____

❺ _____

❻ _____

❼ _____

❽ _____

❾ _____

❿ _____

⓫ _____

⓬ _____

⓭ _____

SUMMARY

Foodhandlers can contaminate food at every step in its flow through the establishment. Good personal hygiene is a critical protective measure against contamination and foodborne illness. A successful personal hygiene program depends on trained foodhandlers who possess the knowledge, skills, and attitude necessary to maintain a safe food system.

Foodhandlers have the potential to contaminate food when they have been diagnosed with a foodborne illness, show symptoms of a gastrointestinal illness, have infected lesions, or when they touch anything that might contaminate their hands. Foodhandlers must pay close attention to what they do with their hands since simple acts such as nose picking or running fingers through the hair can contaminate food. Proper handwashing must always be practiced. This is especially important before starting work, after using the restroom, after sneezing, coughing, smoking, eating, or drinking, and before and after handling raw food. It is up to the manager to monitor handwashing to make sure it is thorough and frequent. In addition, hands need other care to ensure they will not transfer contaminants to food. Fingernails should be kept short and clean. Cuts and sores should be covered with clean bandages. Hand cuts should also be covered with gloves or finger cots.

Gloves can create a barrier between hands and food; however, they should never be used in place of handwashing. Hands must be washed before putting on gloves and when changing to a fresh pair. Gloves used to handle food are for single use and should never be washed or re-used. They must be changed whenever contamination occurs.

All employees must maintain personal cleanliness. They should bathe or shower before work and keep their hair clean. Prior to handling food, foodhandlers must put on a clean hair restraint, clean clothing, and appropriate shoes, and remove jewelry. Aprons should always be removed and properly stored when the employee leaves food-preparation areas.

Establishments should implement strict policies regarding eating, drinking, smoking, and chewing gum and tobacco. These activities should not be allowed when the foodhandler is preparing or serving food or working in food-preparation areas.

Foodhandlers must be encouraged to report health problems to management before working with food. If their condition could contaminate food or equipment, they must stop working and see a doctor. Managers must not allow foodhandlers diagnosed with a foodborne illness to work, and must report illnesses resulting from *Salmonella typhi, Shigella* spp., shiga toxin-producing *E. coli,* and the hepatitis A virus to the local regulatory agency. Managers must restrict foodhandlers from working with or around food if they have symptoms that include fever, diarrhea, vomiting, sore throat with fever, or jaundice. If a foodhandler has any one of these symptoms and the establishment primarily serves a high-risk population, the foodhandler must be excluded from the establishment.

Management plays a critical role in the effectiveness of a personal hygiene program. By establishing a program that includes specific policies, and by training and enforcing those policies, managers can minimize the risk of causing a foodborne illness. Most importantly, managers must set a good example by modeling proper personal hygiene practices.

Apply Your Knowledge

Randall's Day

1 Randall and his manager made several errors. How many can you identify?

▶ If you can identify only eight to twelve errors, you may need to reread this section.

▶ If you can identify thirteen to seventeen errors, you have a good understanding of this section.

▶ If you can identify seventeen or more errors, you are on your way to becoming a health inspector.

For answers, please turn to page 4-23.

Randall is a foodhandler at a deli. It is 7:47 A.M. and Randall has just woken up. He is scheduled to be at work and ready to go by 8:00 A.M. When he gets out of bed, his stomach feels queasy, but he blames that on drinks he consumed the night before.

Fortunately, he lives only five minutes from work, but he does not have enough time to take a shower. He grabs the same uniform he wore the day before when he prepped chicken. Randall is wearing several pieces of jewelry from his night out on the town.

Randall does not have luck on his side today. En route to the restaurant, his oil light comes on and he is forced to pull off the road and add oil to his car. When he walks through the door at work, he realizes he has left his hat at home. Randall is greeted by an angry manager who puts him to work right away, loading the rotisserie with raw chicken. He then moves to serving a customer who orders a freshly made salad. Randall is known for his salads and makes the salad to the customer's approval.

The deli manager, short staffed on this day, asks Randall to take out the garbage, then prepare potato salad for the lunch-hour rush. On the way back in, Randall mentions to the manager that his stomach is bothering him. The manager, thinking of his staff shortage, asks him to stick it out as long as he can. Randall agrees and heads to the restroom in hope of relieving his symptoms. After quickly rinsing his hands in the restroom, he finds that the paper towels have run out. Short of time, he wipes his hands on his apron.

Later, Randall cuts his finger while preparing the potato salad. He bandages the cut and continues his prep work. The manager summons Randall to clean the few tables the deli has made available for customers. Randall puts on a pair of single-use gloves and cleans and sanitizes the tables. When finished, he grabs a piece of chicken from the rotisserie for a snack and immediately goes back to preparing the potato salad because it is almost noon.

Apply Your Knowledge

① Explain how Marty might have caused an outbreak of shigellosis.

② What measures should have been taken to prevent it?

For answers, please turn to page 4-23.

A Case in Point

Marty works for a catering company. A few days ago, he was serving hot food from chafing dishes at an outdoor music festival. He did not wear gloves because he used spoons and tongs to serve the food. His manager noticed that Marty made multiple trips to the bathroom during his four-hour shift. These trips did not interrupt service to customers because there were plenty of staff members on hand and Marty hurried to and from the restroom.

The nearest restroom had soap, separate hot and cold water faucets, and a working hot-air dryer, but no paper towels. Each time Marty used the restroom, he washed his hands quickly and then dried them on his apron. Throughout the following week, the manager of the catering company received several telephone calls from people who had attended the music festival and had eaten their food. They each complained of diarrhea, fever, and chills. One call was from a mother of a young boy who was hospitalized for dehydration. The doctor reported that the boy had shigellosis.

Apply Your Knowledge

Use these questions to test your knowledge of the concepts presented in this section.

Multiple-Choice Study Questions

1. Which of the following personal behaviors can contaminate food?
   A. Touching a pimple
   B. Touching hair
   C. Nose picking
   D. All of the above

2. After you have washed your hands, which of the following items should be used to dry them?
   A. Your apron
   B. A wiping cloth
   C. A common cloth
   D. Single-use paper towels

3. A deli worker stops making sandwiches to use the restroom. She must first
   A. wash her hands.
   B. take off her hat.
   C. take off her apron and properly store it.
   D. change her uniform.

4. Which of the following items can contaminate food?
   A. Rings
   B. A watch
   C. Earrings
   D. All of the above

5. Which of the following is the proper procedure for washing your hands?
   A. Run hot water (at least 100°F [38°C]), moisten hands and apply soap, vigorously scrub hands and arms, apply sanitizer, dry hands.
   B. Run hot water (at least 100°F [38°C]), moisten hands and apply soap, vigorously scrub hands and arms, rinse hands, dry hands.
   C. Run cold water (at least 41°F [5°C]), moisten hands and apply soap, vigorously scrub hands and arms, rinse hands, dry hands.
   D. Run cold water (at least 41°F [5°C]), moisten hands and apply soap, vigorously scrub hands and arms, apply sanitizer, dry hands.

Continued on next page...

Apply Your Knowledge   Multiple-Choice Study Questions *continued*

6. Establishments must only use hand sanitizers that
   A. dry quickly.
   B. can be dispensed in a liquid.
   C. have been approved by the FDA.
   D. can be applied before handwashing.

7. Which foodhandler is least likely to contaminate the food she will handle?
   A. A foodhandler who keeps her fingernails long
   B. A foodhandler who keeps her fingernails short
   C. A foodhandler who wears false fingernails
   D. A foodhandler who wears nail polish

8. Kim wore disposable gloves while she formed raw ground beef into patties. When she was finished, she continued to wear the gloves while she sliced hamburger buns. What mistake did Kim make?
   A. She failed to wash her hands and put on new gloves after handling raw meat and before handling the ready-to-eat buns.
   B. She failed to wash her hands before wearing the same gloves to slice the buns.
   C. She failed to wash and sanitize her gloves before handling the buns.
   D. She failed to wear re-usable gloves.

9. A foodhandler who has been diagnosed with shigellosis should be
   A. told to stay home.
   B. told to wear gloves while working with food.
   C. told to wash his hands every fifteen minutes.
   D. assigned to a nonfoodhandling position until he is feeling better.

10. Managers must report employee illnesses resulting from this pathogen.
   A. *Shigella* spp.   C. *Clostridium perfringens*
   B. *Vibrio vulnificus*   D. *Clostridium botulinuim*

Continued on next page...

11. Some jurisdictions will allow bare-hand contact with cooked and ready-to-eat food if
    A. employees double-wash their hands.
    B. employees keep their fingernails short and clean.
    C. employees use hand sanitizers after properly washing their hands.
    D. the establishment has a verifiable written policy on handwashing procedures.

12. Foodhandlers should be restricted from working with or around food if they are experiencing which of the following symptoms?
    A. Soreness, itching, fatigue
    B. Fever, vomiting, diarrhea
    C. Headache, irritability, thirst
    D. Muscle cramps, insomnia, sweating

13. Which of the following policies should be implemented at establishments?
    A. Employees must not smoke while preparing or serving food.
    B. Employees must not eat while in food-preparation areas.
    C. Employees must not chew gum or tobacco while preparing or serving food.
    D. All of the above

14. Stephanie has a small cut on her finger and is about to prepare chicken salad. How should Stephanie's manager respond to the situation?
    A. Send Stephanie home immediately.
    B. Cover the hand with a glove or finger cot.
    C. Cover the cut with a clean bandage and a glove or finger cot.
    D. Cover the cut with a clean bandage.

Continued on next page...

Apply Your Knowledge   **Multiple-Choice Study Questions** *continued*

15. Hands should be washed after which of the following activities?

    A. Touching your hair   C. Using a handkerchief
    B. Eating   D. All of the above

16. Al, the prep cook at the Great Lakes Senior Citizen Home, called his manager and told her that he had a bad headache, upset stomach, and a sore throat with fever. What is the manager required to do with Al?

    A. Tell him to rest for a couple hours and then come in.

    B. Tell him to go to the doctor and then immediately come to work.

    C. Tell him that he cannot come to work and that he should see a doctor.

    D. Tell him that he can come in for a couple of hours and then go home.

For answers, please turn to page 4-23.

Apply Your Knowledge   Answers

Page	Activity

4-2 Test Your Food Safety Knowledge
1. False   2. True   3. True   4. True   5. True

4-7 Put the Steps in Order
A. 3   B. 1   C. 5   D. 4   E. 2   F. 6

4-13 Exclusion or Restriction?
1. R
2. E
3. E

4-14 What's Wrong with This Picture?

❶ The woman is smoking. ❷ The woman did not remove her jewelry prior to preparing food. ❸ The woman does not have her hair restrained properly. ❹ The woman is wearing nail polish. ❺ The woman has long fingernails or is wearing false nails. ❻ A man is scratching a bandaged cut. ❼ A man has an unrestrained beard. ❽ A man is drinking from an uncovered container in a food-prep area. ❾ A man is sneezing on the cutting board and prep table. ❿ A man is eating in a food-prep area. ⓫ A man who has been cutting raw meat is about to help another man prepare raw vegetables without removing his gloves and washing his hands. ⓬ A man is wiping his gloved hands on his apron. ⓭ A man is ill and should not be working with food.

4-17 Randall's Day

❶ Randall made several errors. Ⓐ Randall failed to bathe or shower before work. Ⓑ Randall wore a dirty uniform to work. Ⓒ Randall dressed prior to coming to work. Ⓓ Randall failed to remove jewelry prior to preparing and serving food. Ⓔ Randall failed to wear a hair restraint. Ⓕ Randall failed to report his illness to the manager. Ⓖ Randall failed to wash his hands before handling the raw chicken. Ⓗ Randall failed to wash his hands after handling the raw chicken. Ⓘ The manager failed to inquire about the symptoms of Randall's illness. If Randall were to report that he had diarrhea, the manager should restrict him from working with or around food. Ⓙ Randall failed to wash his hands properly after taking out the trash. Ⓚ Randall failed to wash his hands properly after using the restroom. Ⓛ Randall failed to dry his hands properly after washing them, recontaminating his hands when he wiped them on his apron. Ⓜ Randall wore his apron into the restroom. Ⓝ The manager failed to ensure that the restroom was adequately stocked with paper towels. Ⓞ Randall failed to inform the manager after cutting his finger. Ⓟ Randall failed to place a finger cot or a single-use glove over the bandaged finger. Ⓠ Randall failed to wash his hands before putting on the single-use gloves. Ⓡ Randall touched the ready-to-eat chicken with his contaminated gloves. Ⓢ Randall was eating chicken while preparing food.

Continued on the next page...

Apply Your Knowledge   Answers *continued*

Page	Activity

4-18   A Case in Point

❶ Marty had shigellosis. Because he was in a hurry, he failed to wash and dry his hands properly. Even though he used tongs to handle the food, Marty must have made direct contact with the food or food-contact surfaces, which resulted in a foodborne-illness outbreak. ❷ Marty should have informed his manager that he was ill. However, his frequent and hurried trips to the restroom should have indicated this to the manager. If, after exploring the problem, the manager found that Marty was suffering from diarrhea, he should have restricted Marty from working with or around food.

4-19   Multiple-Choice Study Questions

1. D	5. B	9. A	13. D
2. D	6. C	10. A	14. C
3. C	7. B	11. D	15. D
4. D	8. A	12. B	16. C

Apply Your Knowledge   Notes

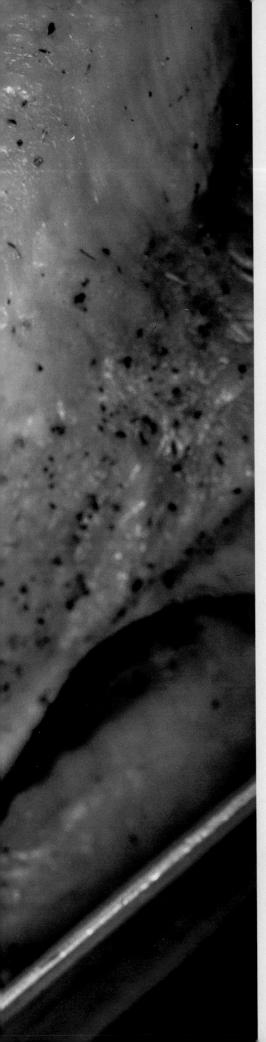

Unit 2

The Flow of Food Through the Operation

The Flow of Food: An Introduction

Inside this section:
▶ Preventing Cross-Contamination
▶ Time and Temperature Control
▶ Monitoring Time and Temperature

After completing this section, you should be able to:
▶ Identify methods for preventing cross-contamination.
▶ Identify methods for preventing time-temperature abuse.
▶ Identify different types of temperature-measuring devices and their uses.

▶ Calibrate and maintain different temperature-measuring devices.
▶ Properly measure the temperature of food at each point in the flow of food.

Apply Your Knowledge	Test Your Food Safety Knowledge

Check to see how much you know about the concepts in this section. Use the page references provided to explore the topic in each question.

❶ **True or False:** The longer food stays at 85°F (29°C), the more time microorganisms have to multiply. *(See page 5-5.)*

❷ **True or False:** The flow of food begins with purchasing and ends with cooking. *(See page 5-3.)*

❸ **True or False:** When checking the temperature of a roast, insert the thermometer stem into the thinnest part of the product. *(See page 5-11.)*

❹ **True or False:** When calibrating a bimetallic stemmed thermometer, it should be set to 32°F (0°C) prior to placing it into ice water. *(See page 5-13.)*

❺ **True or False:** Washing and rinsing a cutting board will prevent it from cross-contaminating the next product it touches. *(See page 5-4.)*

For answers, please turn to page 5-18.

CONCEPTS

▶ **Boiling-point method:** Method of calibrating thermometers based on the boiling point of water.

▶ **Calibration:** Process of ensuring that a thermometer gives accurate readings by adjusting it to a known standard, such as the freezing point or boiling point of water.

▶ **Flow of food:** Path food takes through an establishment, from purchasing and receiving, through storing, preparing, cooking, holding, cooling, reheating, and serving.

▶ **Ice-point method:** Method of calibrating thermometers based on the freezing point of water.

▶ **Thermometer:** Device for accurately measuring the internal temperature of food, the air temperature inside a freezer or cooler, or the temperature of equipment. Bimetallic stemmed thermometers, thermocouples, and thermistors are common types of thermometers used in the restaurant and foodservice industry.

▶ **Time-temperature indicator (TTI):** Time and temperature monitoring device attached to a food shipment to determine if the product's temperature has exceeded safe limits during shipment or later storage.

INTRODUCTION

Your responsibility for the safety of the food in your establishment starts long before you serve meals. Many things can happen to a product on its path through the establishment, from purchasing and receiving, through storing, preparing, cooking, holding, cooling, reheating, and serving—known as the flow of food. (See *Exhibit 5a.*) A frozen product that leaves the processor's plant in good condition, for example, may thaw on its way to the distributor's warehouse and go unnoticed during receiving. Once in your establishment, the product might not be stored properly or cooked to the correct internal temperature, potentially causing a foodborne illness.

The safety of the food you serve at your establishment will depend largely on your understanding of food safety concepts throughout the flow of food, especially the prevention of cross-contamination and time and temperature control. It also depends on your ability to develop a system that prioritizes, monitors, and verifies the most important food safety practices, which will be discussed in Section 10.

PREVENTING CROSS-CONTAMINATION

A major hazard to the flow of food in your operation is cross-contamination, which is the transfer of microorganisms from one food or surface to another. Microorganisms move around easily in a kitchen. They can be transferred from food or unwashed hands to prep tables, equipment, utensils, cutting boards, dish towels, sponges, or other food.

Cross-contamination can occur at almost any point in an operation. When you know where and how microorganisms can be transferred, cross-contamination is fairly simple to prevent. Prevention starts with the creation of barriers between food products. These barriers can be physical or procedural.

Exhibit 5a

The Flow of Food

Purchasing · Receiving · Storing · Preparing · Cooking · Holding · Cooling · Reheating · Serving

The Flow of Food

Physical Barriers for Preventing Cross-Contamination

▶ **Assign specific equipment to each type of food product.** For example, use one set of cutting boards, utensils, and containers for poultry, another set for meat, and a third set for produce. Some manufacturers make colored cutting boards and utensils with colored handles. Color coding can tell employees which equipment to use with what products, such as green for produce, yellow for chicken, and red for meat. Although color coding is helpful, it does not eliminate the need to follow proper practices (i.e., cleaning and sanitizing, minimizing cross-contamination, etc.). Color coding helps to minimize the risk from actually occurring.

▶ **Clean and sanitize all work surfaces, equipment, and utensils after each task.** After cutting up raw chicken, for example, it is not enough to simply rinse the cutting board. Wash, rinse, and sanitize cutting boards and utensils in a three-compartment sink, or run them through a warewashing machine. Make sure employees know which cleaners and sanitizers to use for each job. Sanitizers used on food-contact surfaces must meet local or state department codes conforming to the Code of Federal Regulations (21CFR178.1010). *(See Section 11 for more information on cleaning and sanitizing.)*

Procedural Barriers for Preventing Cross-Contamination

▶ **When using the same prep table, prepare raw and ready-to-eat food at different times.** For example, establishments with limited prep space can prepare lunch salads in the morning, clean and sanitize the utensils and surfaces, and then debone chicken for dinner entrées in the same space in the afternoon.

▶ **Purchase ingredients that require minimal preparation.** For example, an establishment can switch from buying raw chicken breasts to purchasing precooked breasts.

Apply Your Knowledge

Preventing Cross-Contamination

Put an **X** in the space if the practice helps prevent cross-contamination.

____ **❶** Using red cutting boards to prepare meat and green boards to prepare raw vegetables

____ **❷** Washing and rinsing a cutting board after preparing raw fish

____ **❸** Purchasing chopped lettuce rather than chopping it in the establishment

____ **❹** Preparing raw chicken and potato salad on the prep table at the same time

For answers, please turn to page 5-18.

TIME AND TEMPERATURE CONTROL

One of the biggest factors in foodborne-illness outbreaks is time-temperature abuse. Disease-causing microorganisms grow and multiply at temperatures between 41°F and 135°F (5°C and 57°C), which is why this range is known as the temperature danger zone. Microorganisms grow much faster in the middle of the zone, at temperatures between 70°F and 125°F (21°C and 52°C). Whenever food is held in the temperature danger zone, it is being abused.

As you learned in Section 2, time also plays a critical role in food safety. Microorganisms need both time and temperature to grow. The longer food stays in the temperature danger zone, the more time microorganisms have to multiply and make food unsafe. To keep food safe throughout the flow of food, you must minimize the amount of time it spends in the temperature danger zone. It is recommended that food does not remain in the zone for more than four hours.

Common opportunities for time-temperature abuse throughout the flow of food include:

▶ Not cooking food to its required minimum internal temperature

▶ Not cooling food properly

▶ Failing to reheat food to 165°F (74°C) for fifteen seconds within two hours

The Temperature Danger Zone

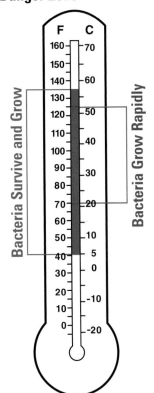

Time and Temperature Control

Print simple forms employees can use to record temperatures.

Reprinted with permission from Roger Bonifield and Dingbats Restaurants

▶ Failing to hold food at a minimum internal temperature of 135°F (57°C) or higher or 41°F (5°C) or lower

The best way to avoid time-temperature abuse is to establish procedures employees must follow and then monitor them. Make time and temperature control part of every employee's job. Some suggestions include:

▶ **Decide the best way to monitor time and temperature in your establishment.** Determine which foods should be monitored, how often, and who should check them. Then assign responsibilities to employees in each area.

Make sure employees understand exactly what you want them to do, how to do it, and why it is important.

▶ **Make sure the establishment has the right kind of thermometers available in the right places.** Give employees their own calibrated thermometers. Have them use timers in prep areas to monitor how long food is being kept in the temperature danger zone.

▶ **Regularly record temperatures and the times they are taken.** Print simple forms employees can use to record temperatures and times throughout the shift. Post these forms on clipboards outside of refrigerators and freezers, near prep tables, and next to cooking and holding equipment. (See *Exhibit 5b.*)

The Flow of Food

Purchasing
Receiving
Storing
Preparing
Cooking
Holding
Cooling
Reheating
Serving

▶ **Incorporate time and temperature controls into standard operating procedures for employees.** These might include:

 ▶ Removing from the refrigerator only the amount of food that can be prepared in a short period of time

 ▶ Refrigerating ingredients and utensils before preparing certain recipes, such as tuna or chicken salad

 ▶ Cooking potentially hazardous food to required minimum internal temperatures

▶ **Develop a set of corrective actions.** Decide what action should be taken if time and temperature standards are not met. For example, an establishment that holds egg rolls on a steam table might throw them out if their internal temperature falls below 135°F (57°C) for more than four hours, or they might reheat the egg rolls to 165°F (74°C) for at least fifteen seconds within two hours.

MONITORING TIME AND TEMPERATURE

To manage both time and temperature, you need to monitor and control them. The thermometer may be the single most important tool you have to protect your food.

Choosing the Right Thermometer

There are many types of thermometers used in an establishment. Each is designed for a specific purpose. Some are used to measure the temperature of refrigerated or frozen storage areas. Others measure the temperature of equipment, such as ovens, hot-holding cabinets, and warewashing machines. Perhaps the most important types are thermometers that measure the temperature of food. The most common types used in establishments are the bimetallic stemmed thermometer, the thermocouple, and the thermistor. Infrared thermometers are also becoming increasingly popular.

Bimetallic Stemmed Thermometer

The most common and versatile type of thermometer used in the restaurant and foodservice industry is the bimetallic stemmed thermometer. This type of thermometer measures temperature through a metal probe with a sensor in the end. Bimetallic

Bimetallic Stemmed Thermometer

Courtesy of Cooper-Atkins Corporation

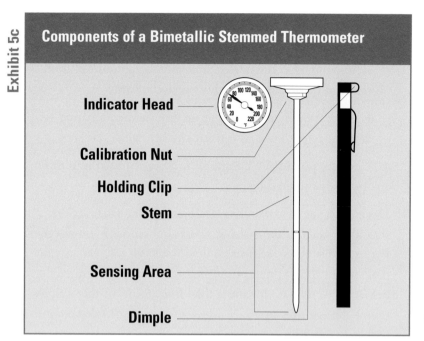

Components of a Bimetallic Stemmed Thermometer

Indicator Head

Calibration Nut

Holding Clip

Stem

Sensing Area

Dimple

stemmed thermometers often have scales measuring temperatures from 0°F to 220°F (–18°C to 104°C). This makes them useful for measuring the temperatures of everything from incoming shipments to the internal temperature of food in hot-holding units. When you select this type of thermometer, it should have (see *Exhibit 5c*)

▶ an adjustable calibration nut to keep it accurate.

▶ easy-to-read, numbered temperature markings.

▶ a dimple to mark the end of the sensing area (which begins at the tip).

▶ accuracy to within ±2°F (±1°C).

Thermocouples and Thermistors

Thermocouples and thermistors measure temperatures through a metal probe or sensing area and display results on a digital readout. They come in a wide variety of styles and sizes, from small pocket models to panel-mounted displays. Many come with interchangeable temperature probes designed to measure the temperature of equipment and food.

Thermocouple

Courtesy of Cooper-Atkins Corporation

Basic types of probes include immersion, surface, penetration, and air probes. (See *Exhibit 5d.*) Immersion probes are designed to measure temperatures of liquids, such as soups, sauces, or frying oil. Surface probes measure temperatures of flat cooking equipment like griddles. Penetration probes are used to measure the internal temperature of food. Air probes measure temperatures inside refrigerators or ovens.

Infrared (Laser) Thermometers

Infrared thermometers use infrared technology to produce accurate temperature readings of food and equipment surfaces.

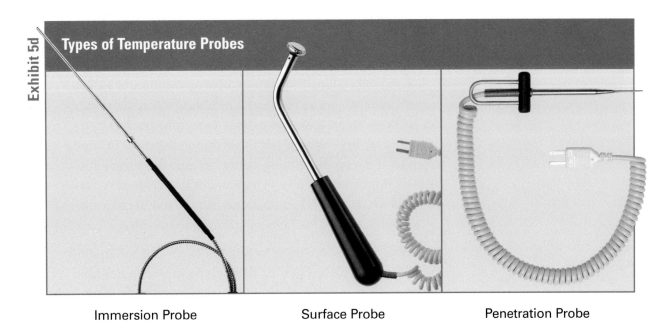

Exhibit 5d

Types of Temperature Probes

Immersion Probe   Surface Probe   Penetration Probe

Courtesy of Cooper-Atkins Corporation

They are quick and easy to use. Infrared thermometers are noncontact thermometers that, when used properly, can reduce the risk of cross-contamination and damage to food products.

When using infrared thermometers, remember the following:

▶ Infrared thermometers should not be used to measure air temperature or the internal temperature of food. They are designed to measure surface temperature.

▶ Hold the thermometer as close as possible to the product without touching it.

▶ Remove any barriers between the thermometer and the product being measured. Do not take temperature measurements through glass or shiny or polished-metal surfaces, such as stainless steel or aluminum.

▶ Always follow the manufacturer's guidelines for tips on obtaining the most accurate temperature reading with the infrared thermometer you are using.

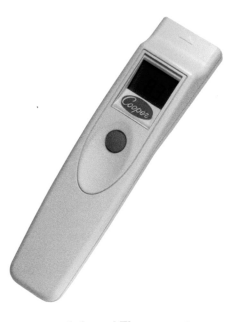

Infrared Thermometer

Courtesy of Cooper-Atkins Corporation

Time-Temperature Indicators (TTI) and Other Time-Temperature Recording Devices

Some instruments are designed to monitor both time and product temperature. The time-temperature indicator (TTI) is one example. Some suppliers attach these self-adhesive tags or sticks to a food shipment to determine if the product's temperature has exceeded safe limits during shipment or later storage. If the product's temperature has exceeded these limits, the TTI provides an irreversible record of the incident. A change in color inside the TTI windows notifies the receiver that the product has undergone time-temperature abuse. (See *Exhibit 5e*.)

More suppliers are using recording devices that continuously monitor temperatures in their delivery trucks. When delivered products appear to have suffered time-temperature abuse, the recording device can be checked to see if the temperature in the delivery truck changed at any time during transit.

Exhibit 5e

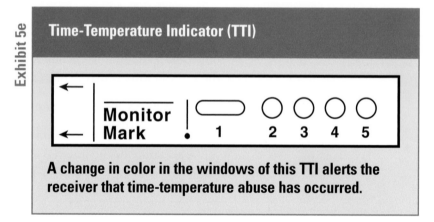

Time-Temperature Indicator (TTI)

A change in color in the windows of this TTI alerts the receiver that time-temperature abuse has occurred.

General Thermometer Guidelines

It is important to know how to use and care for each type of thermometer found in your operation. When it comes to maintenance, always follow manufacturers' recommendations. Here are a few simple guidelines for using thermometers.

▶ **Keep thermometers and their storage cases clean.** Thermometers should be washed, rinsed, sanitized, and air-dried before and after each use to prevent cross-contamination. Use an approved food-contact surface sanitizing solution to sanitize them. Have an adequate supply of clean and sanitized thermometers on hand.

▶ **Calibrate thermometers regularly to ensure accuracy.**
This should be done before each shift or before each day's
deliveries. Thermometers should also be recalibrated any time
they suffer a severe shock (for example, after being dropped or
after an extreme change in temperature). Thermometers that
hang or sit in refrigerators or freezers can be damaged easily.
To make sure these thermometers are accurate, use a
thermocouple with an air temperature probe to check the
temperature. Hanging thermometers usually cannot be
recalibrated and must be replaced if they are not accurate.

▶ **Never use glass thermometers filled with mercury or
spirits to monitor the temperature of food.** They can break
and pose a serious danger to employees and customers.

▶ **Measure internal temperatures of food by inserting the
thermometer stem or probe into the thickest part of the
product (usually the center).** It is a good practice to take at
least two readings in different locations because product
temperatures may vary across the food portion. When checking
the internal temperature of food using a bimetallic stemmed

Apply Your Knowledge	Pick the Right Thermometer

For each situation, select the appropriate temperature measuring device and place the letter in the space provided. Some devices will be selected more than once.	Which temperature-measuring device should be used to check the following?
	 C ❶ Internal temperature of a hamburger patty
	 A ❷ Surface temperature of a steak
	 D ❸ Temperature of a shipment of frozen chicken during transport
	 B ❹ Internal temperature of a roast
	 B ❺ Internal temperature of a large stockpot of soup
	
	     A. Infrared thermometer
	     B. Thermocouple
	     C. Bimetallic stemmed thermometer
	     D. Time-temperature indicator
	
	 For answers, please turn to page 5-18.

thermometer, insert the stem into the product so that it is at least immersed from the tip to the end of the sensing area. When measuring the internal temperature of thin food, such as meat or fish patties, small diameter probes should be used.

▶ **Wait for the thermometer reading to steady before recording the temperature of a food item.** Wait at least fifteen seconds from the time the thermometer stem or probe is inserted into the food.

How to Calibrate Thermometers

Calibration is the process of ensuring that a thermometer gives accurate recordings by adjusting it to a known standard. Most thermometers can be easily calibrated. Two accepted methods of calibration are the boiling-point method (see *Exhibit 5f*) and the ice-point method—which is the most commonly used. (See *Exhibit 5g*.) To calibrate your thermometers properly, follow one of these methods.

	Boiling-Point Method for Calibrating a Thermometer	
Step	**Process**	**Notes**
❶	Bring clean tap water to a boil in a deep pan.	
❷	Put the thermometer stem or probe into the boiling water so the sensing area is completely submerged. Wait thirty seconds, or until the indicator stops moving.	Do not let the stem or probe touch the pan's bottom or sides. The thermometer stem or probe must remain in the boiling water.
❸	Hold the calibration nut securely with a wrench or other tool and rotate the head of the thermometer until it reads 212°F (100°C) or the appropriate boiling-point temperature for your elevation.	The boiling point of water is about 1°F (about 0.5°C) lower for every 550 feet (168 m) you are above sea level. On some thermocouples or thermistors, it may be possible to press a reset button to adjust the readout.

Exhibit 5f

Ice-Point Method for Calibrating a Thermometer

Step	Process		Notes
❶	Fill a large container with crushed ice. Add clean tap water until the container is full.		Stir the mixture well.
❷	Put the thermometer stem or probe into the ice water so the sensing area is completely submerged. Wait thirty seconds, or until the indicator stops moving.		Do not let the stem or probe touch the container's bottom or sides. The thermometer stem or probe must remain in the ice water.
❸	Hold the calibration nut securely with a wrench or other tool and rotate the head of the thermometer until it reads 32°F (0°C).		On some thermocouples or thermistors, it may be possible to press a reset button to adjust the readout.

Apply Your Knowledge	Calibrate the Thermometer
Put the steps for calibrating a bimetallic stemmed thermometer in the proper order by placing the number of the step in the space provided.	___ **Ⓐ** Rotate the head of the thermometer until it reads 32°F (0°C). ___ **Ⓑ** Submerge the sensing area of the thermometer stem or probe, and wait for the reading to steady. ___ **Ⓒ** Fill a container with crushed ice and clean tap water. ___ **Ⓓ** Hold the adjusting nut with a wrench or other tool. **For answers, please turn to page 5-18.**

SUMMARY

The flow of food is the path that food takes through your establishment from purchasing and receiving through storing, preparing, cooking, holding, cooling, reheating, and serving. Many things can happen to food as it flows through the establishment, but hazards that the manager has most control over are cross-contamination and time-temperature abuse.

Cross-contamination is the transfer of microorganisms from one food or surface to another. Prevention starts with the creation of physical or procedural barriers between food products. Physical barriers include assigning specific equipment to each type of food product, and cleaning and sanitizing all work surfaces, equipment, and utensils after each task. Procedural barriers include purchasing ingredients that require minimal preparation and preparing raw and ready-to-eat food at different times.

Food has been time-temperature abused any time it has been allowed to remain at temperatures between 41°F and 135°F (5°C and 57°C). This temperature range is known as the temperature danger zone. To keep food safe, you must minimize the amount of time food spends in this dangerous range. To prevent time-temperature abuse in your establishment, you should incorporate time and temperature controls into your standard operating procedures, make thermometers available to your employees, and regularly record temperatures and the times they were taken.

Thermometers are the most important tools managers have to prevent time-temperature abuse. Managers should make sure employees know what different thermometers are used for and how to calibrate and use them properly. Thermometers should be washed, rinsed, sanitized, and air-dried before and after each use to prevent cross-contamination. They should also be calibrated regularly to ensure accuracy. When measuring the internal temperature of food, the thermometer stem or probe should be inserted in the thickest part of the product (usually the center). When using a bimetallic stemmed thermometer, the stem should be immersed in the product from the tip to the end of the sensing area. When measuring the internal temperature of thin food, use a small diameter probe. Always wait for the thermometer reading to steady before recording the temperature of the food. Never use glass thermometers filled with mercury to measure food temperatures.

Thermometers can be calibrated by using either the ice-point or boiling-point methods. Using the ice-point method, the thermometer is submerged in ice water and adjusted to 32°F (0°C), while the boiling-point method requires the thermometer to be adjusted to 212°F (100°C), or the appropriate boiling point for your elevation, after the stem or probe is placed in boiling water.

Apply Your Knowledge

Use these questions to test your knowledge of the concepts presented in this section.

Multiple-Choice Study Questions

1. An employee has just trimmed raw chicken on a cutting board and must now use the board to prepare vegetables. What should the employee do with the board prior to preparing the vegetables?

    A. Wash, rinse, and sanitize the cutting board.
    B. Dry it with a paper towel.
    C. Rinse it under very hot water.
    D. Turn it over and use the reverse side.

2. All of the following practices can help prevent cross-contamination during food preparation *except*

    A. preparing meat separately from ready-to-eat food.
    B. assigning specific equipment for preparing specific food.
    C. rinsing cutting boards between preparing raw food and ready-to-eat food.
    D. using specific storage containers for specific food.

3. Infrared thermometers should be used to measure the

    A. air temperature of a refrigerator.
    B. internal temperature of a cooked turkey.
    C. surface temperature of a steak.
    D. internal temperature of a batch of soup.

4. All of the following practices can help prevent time-temperature abuse *except*

    A. storing milk at 41°F (5°C).
    B. holding chicken noodle soup at 120°F (49°C).
    C. reheating chili to 165°F (74°C) for fifteen seconds within two hours.
    D. holding the ingredients for tuna salad at 39°F (4°C).

5. You have a thermocouple with several different types of probes. Which probe should you use to check the temperature of a large stockpot of soup?

    A. Penetration probe   C. Air probe
    B. Surface probe   D. Immersion probe

Continued on next page...

6. Which step for calibrating a thermometer using the ice-point method is *incorrect*?
   A. First, insert the thermometer stem or probe into a container of ice water.
   B. Second, wait thirty seconds or until the temperature indicator stops moving.
   C. Third, remove the thermometer stem or probe from the ice water.
   D. Finally, adjust the thermometer until it reads 32°F (0°C).

7. Your manager has asked you to purchase a new thermometer for the restaurant. Which type would *not* be a proper choice?
   A. Bimetallic stemmed thermometer accurate to ±2°F (±1°C)
   B. Thermistor
   C. Thermocouple
   D. Mercury-filled glass thermometer

For answers, please turn to page 5-18.

Apply Your Knowledge   Answers

Page	Activity			

5-2   Test Your Food Safety Knowledge
1. True   2. False   3. False   4. False   5. False

5-5   Preventing Cross-Contamination
1, 3

5-11   Pick the Right Thermometer
1. B   3. D   5. B
2. A   4. B, C

5-14   Calibrate the Thermometer
A. 4   C. 1
B. 2   D. 3

5-16   Multiple-Choice Study Questions
1. A   4. B   7. D
2. C   5. D
3. C   6. C

Apply Your Knowledge   Notes

6

The Flow of Food: Purchasing and Receiving

Inside this section:
▶ General Purchasing and Receiving Principles
▶ Receiving and Inspecting Food

After completing this section, you should be able to:
▶ Identify an approved food source.
▶ Identify accept and reject criteria for:
  ▷ Meat and poultry
  ▷ Seafood
  ▷ Milk and dairy products
  ▷ Eggs
  ▷ Fruit and vegetables
  ▷ Canned goods and other dry food
  ▷ Ready-to-eat food
  ▷ Frozen food
  ▷ Bakery goods

Apply Your Knowledge	**Test Your Food Safety Knowledge**
Check to see how much you know about the concepts in this section. Use the page references provided to explore the topic in each question.	① **True or False:** Upon arrival, a delivery of fresh fish should be received at an internal temperature of 41°F (5°C) or lower. *(See page 6-7.)*

❷ **True or False:** Turkey should be rejected if the texture is firm and springs back when touched. *(See page 6-6.)*

❸ **True or False:** You should reject a delivery of frozen steaks covered in large ice crystals. *(See page 6-13.)*

❹ **True or False:** If a sack of flour is dry upon delivery, the contents may still be contaminated. *(See page 6-16.)*

❺ **True or False:** A supplier that is in compliance with local, state, and federal law can be considered an approved source. *(See page 6-3.)*

For answers, please turn to page 6-26.

CONCEPTS

The Flow of Food

▶ **MAP food:** MAP stands for Modified Atmosphere Packaging, a packaging process by which air is removed from a food package and replaced with gases, such as carbon dioxide and nitrogen. These gases help extend the product's shelf life.

▶ ***Sous vide* food:** Food processed by this method is vacuum-packed in individual pouches, partially or fully cooked, and then chilled. This food is then heated for service in the establishment.

▶ **UHT (ultra-high temperature) food:** UHT food is heat-treated at very high temperatures (pasteurized) for a short time to kill microorganisms. These foods are often also aseptically packaged—sealed under sterile conditions to keep them from being contaminated.

▶ **Shellstock tags:** Each container of live, molluscan shellfish received must have an ID tag, on which the delivery date must be written. Tags are to be kept on file for ninety days after the last shellfish is used.

GENERAL PURCHASING AND RECEIVING PRINCIPLES

▶ **Buy only from suppliers who get their products from approved sources.** An approved food source is one that has been inspected and is in compliance with applicable local, state, and federal law. Before you accept any deliveries, it is your responsibility to ensure that food you purchase comes from suppliers (distributors) and sources (points of origin) that have been approved.

▶ **Make sure suppliers are reputable.** Ask other operators what their experience has been with a particular supplier.

▶ **Schedule deliveries for off-peak hours and receive only one delivery at a time.** Arrange it so products are delivered when employees have adequate time to inspect them.

▶ **Make sure enough trained staff are available to promptly receive, inspect, and store food.** They should be authorized to accept, reject, and sign for deliveries.

▶ **Inspect deliveries carefully.** Check for proper labeling, temperature, appearance, and other factors important to safety.

▶ **Use properly calibrated thermometers to sample temperatures of received food items.**

▶ **Check shipments for intact packaging and signs of refreezing, prior wetness, and pest infestation.** Broken boxes, leaky packages, or dented cans are signs of mishandling and could be grounds for rejecting the shipment.

▶ **Inspect deliveries immediately and put items away as quickly as possible.** This is especially true for refrigerated and frozen products.

RECEIVING AND INSPECTING FOOD

Food delivered to your establishment should be inspected carefully. Internal temperatures should be checked and recorded. (See *Exhibit 6a* on the next page.) Other conditions should be checked as well, such as color, texture, odor, and packaging.

Exhibit 6a

Checking the Temperatures of Various Types of Food

Item	Method	Example
Meat, Poultry, Fish	Insert the thermometer stem or probe directly into the thickest part of the product (usually the center).	
MAP, Vacuum-Packed, and *Sous Vide* Food (refrigerated and frozen)	Insert the thermometer stem or probe between two packages, being careful not to puncture them.	
Liquids or Other Packaged Food	Open the package and insert the thermometer stem or probe into the food until the sensing area is immersed. Do not let the thermometer stem or probe touch the sides or bottom of the container.	
Bulk Liquids	Fold the bag around the thermometer stem or probe.	

USDA Inspection Stamp

USDA Grading Stamp

Meat

Meat must be purchased from plants inspected by the USDA or state department of agriculture. Inspected meat will contain a mandatory inspection stamp on packaging and on inspected carcasses. *Inspected* does not mean the product is free of microorganisms but that the product and processing plant have met certain standards.

Most meat and poultry also carry a stamp indicating its grade or palatability, and level of quality. Grading is a voluntary service provided by the USDA and is paid for by processors and packers. USDA grades are printed inside a shield-shaped stamp. Note the examples of inspection and grading stamps to the left. See *Exhibit 6b* for specific *Accept* and *Reject* criteria for deliveries of fresh meat.

Exhibit 6b

Receiving Criteria for Meat	
Accept	**Reject**
Temperature: 41°F (5°C) or lower	**Temperature:** >41°F (5°C)
Color:	**Color:**
▶ Beef: bright, cherry red; aged beef may be darker in color; vacuum-packed beef will appear purplish in color	▶ Beef: brown or green
▶ Lamb: light red	▶ Lamb: brown, whitish surface covering the lean meat
▶ Pork: light pink meat, firm white fat	▶ Pork: excessively dark color, soft or rancid fat
Texture: firm and springs back when touched	**Texture:** slimy, sticky, or dry
Odor: no odor	**Odor:** sour odor
Packaging: intact and clean	**Packaging:** broken cartons, dirty wrappers, torn packaging, vacuum packaging with broken seals

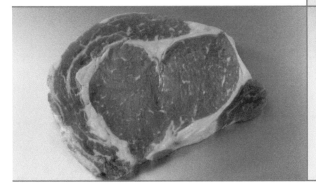

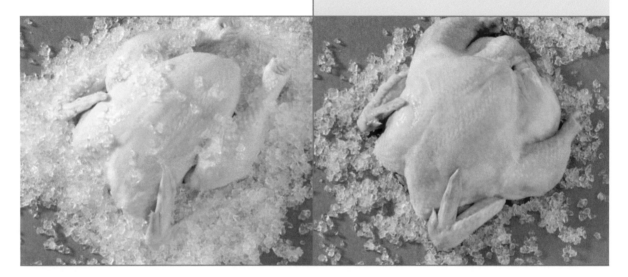

Exhibit 6c

Receiving Criteria for Poultry	
Accept	**Reject**
Temperature: 41°F (5°C) or lower	**Temperature:** >41°F (5°C)
Color: no discoloration	**Color:** purple or green discoloration around the neck; dark wing tips (red tips are acceptable)
Texture: firm and springs back when touched	**Texture:** stickiness under the wings or around joints
Odor: no odor	**Odor:** abnormal, unpleasant odor
Packaging: product should be surrounded by crushed, self-draining ice	

Poultry

USDA Inspection Stamp

USDA Grading Stamp

Poultry is inspected by the USDA or state department of agriculture in much the same way as meat. As with meat, grading is voluntary and paid for by processors. Note the examples of inspection and grading stamps to the left. Fresh poultry should be shipped in self-draining crushed ice, or chill packed. See *Exhibit 6c* for specific *Accept* and *Reject* criteria for deliveries of fresh poultry.

Fish

Fresh fish is very sensitive to time-temperature abuse and can deteriorate quickly if handled improperly. See *Exhibit 6d* for specific *Accept* and *Reject* criteria for deliveries of fresh fish.

Receiving Criteria for Fish

Accept	Reject
Temperature: 41°F (5°C) or lower	**Temperature:** >41°F (5°C)
Color: bright red gills; bright shiny skin	**Color:** dull gray gills; dull dry skin
Texture: firm flesh that springs back when touched	**Texture:** soft flesh that leaves an imprint when touched
Odor: mild ocean or seaweed smell	**Odor:** strong fishy or ammonia smell
Eyes: bright, clear, and full	**Eyes:** cloudy, red-rimmed, sunken
Packaging: product should be surrounded by crushed, self-draining ice	

Shellfish

Shellfish include mollusks such as clams, oysters, and mussels. They can be shipped live, frozen, in the shell, or shucked. Interstate shipping is monitored by the FDA and the Interstate Shellfish Sanitation Conference. Shellfish must be purchased from suppliers listed in the *National Shellfish Sanitation Guide for the Control of Molluscan Shellfish,* or from sources included in the Interstate Certified Shellfish Shippers List.

Live shellfish must be received with shellstock identification tags, which must remain attached to the container they were delivered in until all of the shellfish have been used. Different batches of shellfish must not be mixed. Operators must write the date of delivery on the tags, and keep the tags on file for ninety days from the date the last shellfish was used. See *Exhibit 6e* for specific *Accept* and *Reject* criteria for deliveries of shellfish.

Exhibit 6e

Receiving Criteria for Shellfish

Accept	Reject
Temperature: ▶ Live: receive on ice or at an air temperature of 45°F (7°C) or lower ▶ Shucked: receive at an internal temperature of 45°F (7°C) or lower **Odor:** mild ocean or seaweed smell **Shells:** closed and unbroken (indicates shellfish are alive) **Condition:** if fresh, they are received alive	**Temperature:** ▶ Live: air temperature >45°F (7°C) ▶ Shucked: internal temperature >45°F (7°C) **Texture:** slimy, sticky, or dry **Odor:** strong fishy smell **Shells:** broken shells; open shells that do not close when tapped **Condition:** dead on arrival

Crustacea

Crustacea include shrimp, crab, and lobster. Live lobsters and crabs must be received alive. Those showing weak signs of life should be cooked right away, while dead ones must be discarded or returned to the vendor for credit. See *Exhibit 6f* for specific *Accept* and *Reject* criteria for deliveries of crustacea.

Exhibit 6f

Receiving Criteria for Crustacea

Accept	Reject
Temperature: ▶ **Live:** must be received alive ▶ **Processed:** internal temperature of 41˚F (5˚C) or lower **Odor:** mild ocean or seaweed smell **Shells:** hard and heavy for lobsters and crabs **Condition:** shipped alive; packed with seaweed and kept moist 	**Temperature:** ▶ **Processed:** internal temperature >41˚F (5˚C) **Odor:** strong fishy smell **Shells:** soft **Condition:** dead on arrival (tail fails to curl when lobster is picked up) 

USDA Inspection Stamp

USDA Grading Stamp

Shell Eggs

Purchase shell eggs from approved, government-inspected suppliers. The mandatory USDA inspection stamp on egg cartons indicates federal regulations are enforced to maintain quality and reduce contamination. As with meat and poultry, grading is voluntary and is provided by the USDA. The grading stamp certifies that eggs have been graded for quality under federal and/or state supervision. Note the examples of inspection and grading stamps for eggs to the left.

You should choose suppliers who can deliver eggs within a few days of the packing date. Eggs must be delivered in refrigerated trucks capable of documenting air temperature during transport. Liquid, frozen, and dehydrated eggs must be pasteurized, as required by law, and bear the USDA inspection mark. When delivered, they should be refrigerated or frozen at the proper temperature. Cases or cartons of shell eggs for direct sale to the consumer must display safe-handling instructions on them. See *Exhibit 6g* for specific *Accept* and *Reject* criteria for deliveries of shell eggs.

Exhibit 6g

Receiving Criteria for Shell Eggs

Accept	Reject
Temperature: receive at an air temperature of 45°F (7°C) or lower **Odor:** no odor **Shells:** clean and unbroken	**Temperature:** air temperatures >45°F (7°C) **Odor:** sulfur smell or off odor **Shells:** dirty or cracked

Dairy

Purchase only pasteurized dairy products since unpasteurized products are potential sources of foodborne pathogens such as *Listeria monocytogenes* and *Salmonella* spp. All milk and milk products should be labeled Grade A, which means they meet standards for quality and sanitary processing methods set by the FDA and the U.S. Public Health Service. Dairy products with the Grade A label—such as cream, cottage cheese, butter, and ice cream—are made with pasteurized milk. See *Exhibit 6h* for specific *Accept* and *Reject* criteria for dairy deliveries.

Exhibit 6h

Receiving Criteria for Dairy

Accept	Reject
Temperature: 41°F (5°C) or lower unless otherwise specified by law	**Temperature:** >41°F (5°C), unless otherwise specified
Milk: sweetish flavor	**Milk:** sour, bitter, or moldy taste
Butter: sweet flavor, uniform color, firm texture	**Butter:** sour, bitter, or moldy taste; uneven color; soft texture
Cheese: typical flavor, texture, and uniform color	**Cheese:** abnormal flavor or texture, uneven color, unnatural mold

Produce

Fresh fruit and vegetables have different temperature requirements for transportation and storage. No specific temperature is mandated by regulation with the exception of cut melons, which must be received and stored at 41°F (5°C) or lower. Products that contain evidence of pests should always be rejected. See *Exhibit 6i* for specific *Accept* and *Reject* criteria for produce deliveries.

Exhibit 6i

Receiving Criteria for Fresh Produce	
Accept	**Reject**
Conditions: vary according to produce item; only accept items that show no sign of spoilage 	**Conditions:** grounds for rejecting one produce item may not apply to another; signs of spoilage include: ▶ Insect infestation ▶ Mold ▶ Cuts ▶ Mushiness ▶ Discoloration ▶ Wilting ▶ Dull appearance ▶ Unpleasant odors and tastes 

Refrigerated and Frozen Processed Food

More and more establishments are using prepared food that is either refrigerated or frozen. This includes pre-cut meats, refrigerated or frozen entrées that only require heating, and fresh-cut fruit and vegetables. See *Exhibit 6j* for specific *Accept* and *Reject* criteria for refrigerated or frozen processed food.

Exhibit 6j

Receiving Criteria for Refrigerated and Frozen Processed Food

Accept	Reject
Refrigerated Food ▶ **Temperature:** 41°F (5°C) or lower unless specified by the manufacturer ▶ **Packaging:** intact and in good condition	**Refrigerated Food** ▶ **Temperature:** >41°F (5°C) unless otherwise specified ▶ **Packaging:** torn packages or packages with holes; expired product use-by dates
Frozen Food ▶ **Temperature:** frozen food should be received frozen; ice cream should be received at 6°F to 10°F (−14°C to −12°C) ▶ **Packaging:** intact and in good condition; dry	**Frozen Food** ▶ **Temperature:** food that is not frozen; ice cream at temperatures >6°F to 10°F (−14°C to −12°C) ▶ **Packaging:** torn packages or packages with holes; fluids or frozen liquids in cases, ice crystals or water stains on packaging (evidence of thawing and refreezing) ▶ **Product:** large ice crystals on product (evidence of thawing and refreezing)

MAP, Vacuum-Packed, and *Sous Vide* Packaged Food

MAP stands for modified atmosphere packaging. By this method, air is removed from a food package and replaced with gases, such as carbon dioxide and nitrogen, which help extend the shelf-life of the product. Many fresh-cut produce items are packaged this way.

Vacuum-packed food, such as bacon, is processed by removing the air around the product sealed in a package. Food packaged by the *sous vide* method is vacuum-packed in individual pouches, partially or fully cooked, and then chilled. This food is then heated for service in the establishment. Some frozen, pre-cooked meals are packaged this way. See *Exhibit 6k* for specific *Accept* and *Reject* criteria for MAP, vacuum-packed, and *sous vide* packaged food deliveries.

Exhibit 6k

Receiving Criteria for MAP, Vacuum-Packed, and *Sous Vide* Packaged Food

Accept	Reject
Temperature: ▶ Refrigerated: 41°F (5°C) or lower unless specified by the manufacturer ▶ Frozen: received frozen **Packaging:** intact and in good condition; valid code dates **Product:** acceptable color	**Temperature:** ▶ Refrigerated: >41°F (5°C), unless otherwise specified ▶ Frozen: product is not frozen **Packaging:** torn or leaking packages; expired code dates **Product:** unacceptable color; slime or bubbles

Canned Food

Canned food must be checked carefully for damage. Look for exterior damage and spot-check the contents. Never taste canned product you are unsure of, since consuming *Clostridium botulinum* may be fatal. See *Exhibit 6l* for specific *Accept* and *Reject* criteria for canned food deliveries.

Exhibit 6l

Receiving Criteria for Canned Food

Accept	Reject
Can: can and seal are in good condition **Product:** normal color, texture, odor	**Can:** swollen ends, leaks and flawed seals, rust, dents, no labels **Product:** foamy, milky, or has an abnormal color, texture, or odor

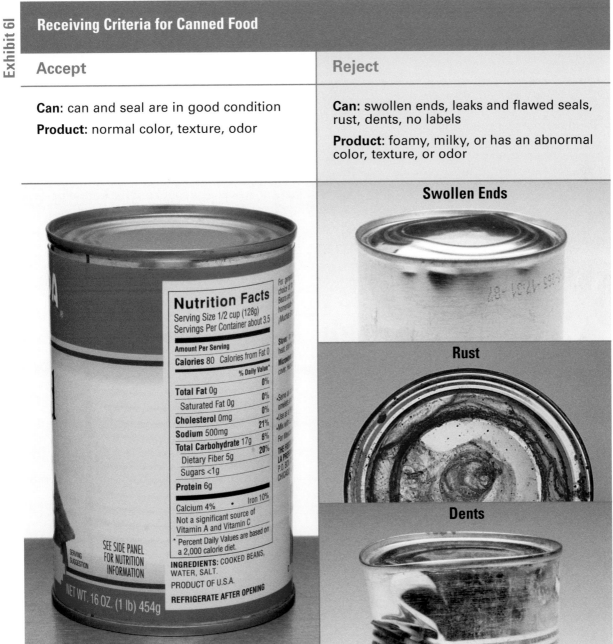

Swollen Ends

Rust

Dents

Dry Food

Dry food must be kept dry. Most microorganisms need moisture to grow and multiply, which is why dry food has a much longer shelf life than fresh food. Often it attracts pests. See *Exhibit 6m* for specific *Accept* and *Reject* criteria for dry food deliveries.

Exhibit 6m

Receiving Criteria for Dry Food	
Accept	**Reject**
Packaging: intact and in good condition **Product:** normal color and odor	**Packaging:** holes, tears, or punctures; dampness or water stains on outer cases and inner packaging (indicates it has been wet) **Product:** abnormal color or odor; spots of mold, or slimy appearance; contains insects, insect eggs, or rodent droppings

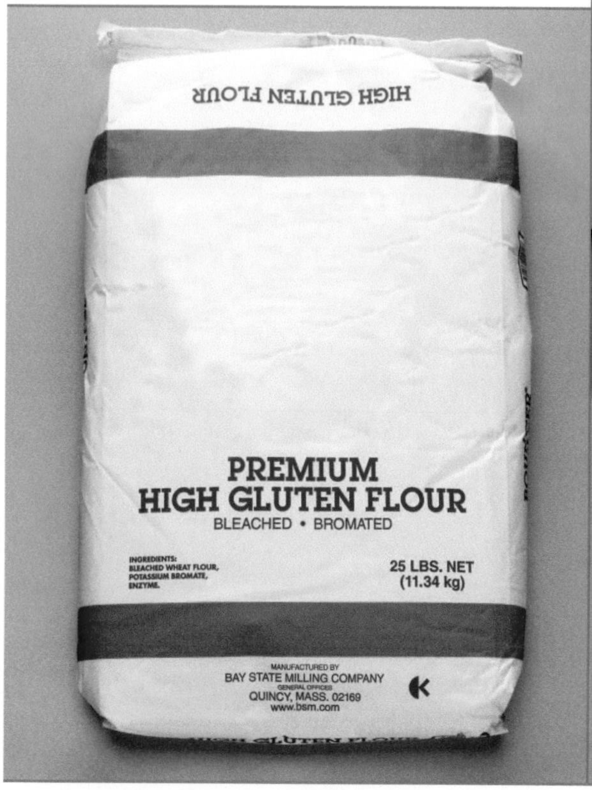

Ultra-High Temperature (UHT) Pasteurized and Aseptically Packaged Food

Some food is heat-treated at very high temperatures to kill microorganisms in a process called ultra-high temperature (UHT) pasteurization. This food is often also aseptically packaged—sealed under sterile conditions to keep it from being contaminated. Examples include some puddings, juices, and creamers and milk products.

Once food has been UHT-pasteurized and aseptically packaged, it can be received and stored at room temperature. Once opened, however, it should be refrigerated at 41°F (5°C) or lower. Always follow manufacturers' directions for receiving and storing these products. See *Exhibit 6n* for specific *Accept* and *Reject* criteria for UHT food deliveries.

Exhibit 6n

Receiving Criteria for Ultra-High Temperature (UHT) Pasteurized and Aseptically Packaged Food

Accept	Reject
Temperature: ▶ UHT food aseptically packaged: room temperature ▶ UHT food not aseptically packaged: follow manufacturer's directions or 41°F (5°C) or lower **Packaging:** intact packaging and seals	**Temperature:** ▶ UHT food not aseptically packaged: >41°F (5°C) **Packaging:** punctured packaging or broken seals

Bakery Goods

It is common for establishments to regularly receive shipments of various types of bakery items. While safe handling depends upon the items received, it is always important to follow manufacturers' recommendations, especially regarding time and temperature control. See *Exhibit 6o* for specific *Accept* and *Reject* criteria for deliveries of bakery goods.

Exhibit 6o

Receiving Criteria for Bakery Goods	
Accept	**Reject**
Temperature: receive at the temperature specified by the manufacturer **Packaging:** intact 	**Temperature:** temperatures higher than those specified by the manufacturer **Packaging:** torn packaging, signs of pest damage **Product:** signs of pest damage, mold 

Potentially Hazardous Hot Food

Occasionally operators may receive a shipment of potentially hazardous, hot food. It must be properly cooked as required by local or federal codes. If you purchase hot food, make sure suppliers have a HACCP plan or other means of documenting proper cooking methods and temperatures. See *Exhibit 6p* for specific *Accept* and *Reject* criteria for deliveries of potentially hazardous hot food.

Exhibit 6p

Receiving Criteria for Potentially Hazardous Hot Food

Accept	Reject
Temperature: 135°F (57°C) or higher **Container:** able to maintain proper temperatures	**Temperature:** <135°F (57°C) **Container:** unable to maintain temperatures

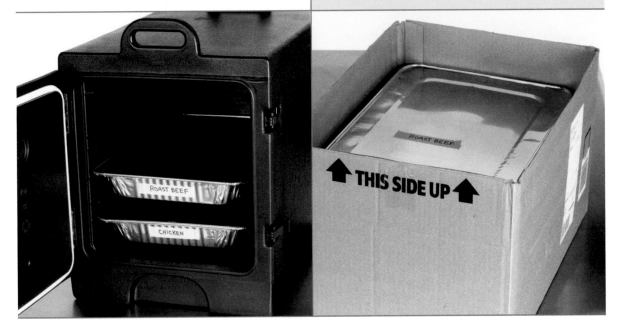

Apply Your Knowledge	Accept It or Reject It

Place an **A** next to each statement if you should accept the product or an **R** if you should reject it.

___ ❶ Beef roasts that are bright red

___ ❷ Chicken received at an internal temperature of 50°F (10°C)

___ ❸ Eggs received at an air temperature of 45°F (7°C)

___ ❹ Fresh salmon with flesh that springs back when touched

___ ❺ Flour that is damp

___ ❻ Processed lobster received at an internal temperature of 45°F (7°C)

___ ❼ Live oysters that have a mild seaweed smell

___ ❽ Frozen meat with large ice crystals on the meat and package

___ ❾ Clams with shells that do not open when tapped

___ ❿ Fresh turkey with dark wing tips

For answers, please turn to page 6-26.

SUMMARY

Even though federal and state agencies regulate and monitor the production and transportation of food such as meat, poultry, seafood, eggs, dairy products, and canned goods, it is your responsibility to check the quality and safety of food that comes into your establishment.

Make sure suppliers are getting their products from approved sources—those that have been inspected and are in compliance with local, state, and federal law. Take steps to ensure that the suppliers you have chosen are reputable by asking other operators what their experiences have been with those suppliers.

Operators must plan delivery schedules so products can be handled promptly and correctly. Employees assigned to receive deliveries should be trained to inspect food properly, as well as to distinguish between products that are acceptable and those that are not. They should also be authorized to reject products that do not meet company standards and to sign for products that do.

The Flow of Food

All products arriving at the establishment should meet agreed-upon standards. Packaging should be clean and undamaged. Use-by dates should be current. Food should show no signs of mishandling.

Products must be delivered at the proper temperature. All products—especially meat, poultry, and fish—should be checked for proper color, texture, and odor. Live, molluscan shellfish and crustacea must be delivered alive. Eggs should be inspected for freshness and for dirty and cracked shells. Dairy products must be checked for freshness. Produce should be fresh and wholesome. Frozen food should be inspected for signs of thawing and refreezing. MAP, vacuum-packed, and *sous vide* food should not bubble or appear slimy, its packaging should be intact, and code dates should not be expired. Canned food must be carefully examined for signs of damage. Dry food should be inspected for pest infestation and moisture. Bakery goods should not be moldy, show signs of pest damage, and should not have passed their expiration date. Potentially hazardous hot food must be delivered at 135°F (57°C) or higher.

Apply Your Knowledge

A Case in Point

❶ What was done wrong?

❷ What could be the result?

For answers, please turn to page 6-26.

ABC Seafood makes its usual Thursday afternoon delivery to The Fish House. John, a prep cook, is the only person in the kitchen when the driver rings the bell at the back door. The kitchen manager, who is in charge of receiving, is in a managers' meeting. The chef is out on an errand, and the rest of the kitchen staff is on break, though some are still in the restaurant.

John follows the driver onto the dock, where the driver unloads two crates of ice-packed fresh fish, bags of live mussels, two buckets of live oysters, a case of shucked oysters in plastic containers, and a case each of frozen shrimp and frozen lobster tails. John goes back into the kitchen to get a bimetallic stemmed thermometer and remembers to look in the chef's office for a copy of the order form. He takes both out to the dock and begins to inspect the shipment.

John checks the products against both the order sheet and the invoice, then begins to check product temperatures. First he checks the temperature of the shucked oysters by taking the cover off one container and inserting the thermometer stem into it. The thermometer reads 45°F (7°C), which worries him. He remembers being taught that refrigerated products should be 41°F (5°C) or lower. He decides not to say anything.

After wiping the thermometer stem on his apron, John checks the internal temperature of a whole fish packed in ice. Finally, he reaches into one of the buckets of live oysters with his hand to see if it feels cold. He notices a few mussels and oysters with open and broken shells. He removes those with broken shells, knowing the chef will not use them.

John records all his findings on the order sheet he took from the chef's office, signs for the delivery, and starts putting the products away. He first puts away the live shellfish, dumping the few still left in the refrigerator into the new containers to make room. Then he puts the shucked oysters and fresh fish in the refrigerator and the shrimp and lobster into the freezer.

Apply Your Knowledge

Multiple-Choice Study Questions

Use these questions to test your knowledge of the concepts presented in this section.

1. Which is the most important factor in choosing a food supplier?
   A. Its prices are the lowest.
   B. Its warehouse is close to your establishment.
   C. It offers a convenient delivery schedule.
   D. It has been inspected and is compliant with local, state, and federal law.

2. Which of the following shipments should be rejected?
   A. Beef that is bright, cherry red in color
   B. Lamb with flesh that is firm and springs back when touched
   C. Fish that arrives with sunken eyes
   D. Chicken received at 41°F (5°C)

3. How should whole, fresh salmon be packaged for delivery and storage?
   A. Layered with salt
   B. Vacuum-sealed
   C. Wrapped in dry, clean cloth
   D. Packed in self-draining, crushed ice

4. Which condition would cause you to reject a shipment of live oysters?
   A. Most of the oysters have closed shells.
   B. The oysters have a mild seaweed scent.
   C. Most of the oysters shells are open and do not close when they are tapped.
   D. The oysters are delivered with shellstock identification tags.

Continued on next page...

Apply Your Knowledge   **Multiple-Choice Study Questions** *continued*

5. A box of sirloin steaks carries both a USDA inspection stamp and a USDA-choice grading stamp. What do these stamps tell you?

   A. The farm that supplied the beef uses only USDA-certified animal feed.

   B. The meat and processing plant have met USDA standards, and the meat quality is acceptable.

   C. The meat wholesaler meets USDA quality-grading standards.

   D. The steaks are free of disease-causing microorganisms.

6. Which condition would cause you to reject a shipment of fresh chicken?

   A. The flesh is firm and springs back when touched.

   B. The wing tips are brown.

   C. The grading stamp is missing.

   D. The chicken is odorless.

7. Statements from a dairy supplier's sales brochure are listed below. Which statement should tell you not to hire this supplier?

   A. We make our cheese with only the freshest unpasteurized milk.

   B. From farm to you, our milk is kept at temperatures below 41°F (5°C).

   C. Money-back guarantee—our prices are the lowest in the area.

   D. We deliver according to your schedule and needs.

8. A shipment of eggs should be rejected for all of the following reasons *except*

   A. the shells are cracked.

   B. they have a sulfur smell.

   C. they lack an inspection stamp.

   D. the air temperature of the delivery truck was 45°F (7°C).

Continued on next page...

9. Which of the following food does not have to be received at 41°F (5°C) or lower?
   A. Beef
   B. Live shellfish
   C. Pork
   D. Fish

10. All of the following would be grounds for rejecting a case of frozen food *except*
    A. there are large ice crystals on the frozen food inside the case.
    B. the outside of the case is water-stained.
    C. the food in the box is frozen solid.
    D. there is frozen liquid at the bottom of the case.

11. Which delivery should be rejected?
    A. Several cans in a case of peaches have torn labels.
    B. Several cans in a case of tomato soup have swollen ends.
    C. A bag of oatmeal is delivered at 60°F (16°C).
    D. A case of rice is missing a USDA inspection stamp.

12. You have just received a shipment of pumpkin pies. What would be grounds for rejecting the shipment?
    A. The pies have been received frozen.
    B. The pies have passed their expiration date.
    C. The pies contain preservatives.
    D. The pies have been received at 41°F (5°C).

For answers, please turn to page 6-26.

Apply Your Knowledge   Answers

Page	Activity

6-2   Test Your Food Safety Knowledge

1. True   2. False   3. True   4. True   5. True

6-20   Accept It or Reject It

1. A   6. R
2. R   7. A
3. A   8. R
4. A   9. A
5. R   10. R

6-22   A Case in Point

❶ John had good intentions and did most things correctly. However, he did make mistakes. John should have:Ⓐ Notified the kitchen manager that the shipment had arrived.Ⓑ Made sure that the bimetallic stemmed thermometer he took from the kitchen had been calibrated properly, as well as clean and sanitized, before using it.Ⓒ Cleaned and sanitized the thermometer after checking the temperature of each product. He should not have wiped the thermometer on his apron.Ⓓ Inserted the thermometer stem into the middle of the bucket of live oysters between the shellfish for an ambient reading instead of trying to judge how cold they were with his hand. The temperature of the oysters should have been 45°F (7°C) or lower. Ⓔ Marked the delivery date on the shellstock identification tags attached to the shipment of oysters and mussels. He should not have dumped the remaining shellfish from the previous shipment into the new containers, mixing one shipment with another. ❷ A foodborne illness could have occurred because of the mistakes John made.

6-23   Multiple-Choice Study Questions

1. D   5. B   9. B
2. C   6. B   10. C
3. D   7. A   11. B
4. C   8. D   12. B

Apply Your Knowledge   **Notes**

7

The Flow of Food: Storage

Inside this section:

▶ General Storage Guidelines
▶ Refrigerated Storage
▶ Frozen Storage

▶ Dry Storage
▶ Storing Specific Food

After completing this section, you should be able to:

▶ Properly label and date-mark refrigerated, frozen, and dry food prior to storage.

▶ Properly store refrigerated, frozen, dry, and canned food.

▶ Apply first in, first out (FIFO) practices as they relate to refrigerated, frozen, and dry-storage areas.

▶ Properly store raw food to prevent cross-contamination.

▶ Identify temperature requirements for refrigerated and dry storage areas.

▶ Identify proper storage containers for refrigerated, frozen, and dry food.

Apply Your Knowledge

Check to see how much you know about the concepts in this section. Use the page references provided to explore the topic in each question.

Test Your Food Safety Knowledge

❶ **True or False:** Potato salad that has been prepared in-house and stored at 41°F (5°C) must be discarded after three days. *(See page 7-4.)*

❷ **True or False:** Food can be stored near chemicals as long as the chemicals are stored in sturdy, clearly labeled containers. *(See page 7-4.)*

❸ **True or False:** Storing cans of stewed tomatoes at 65°F (18°C) is acceptable. *(See page 7-11.)*

❹ **True or False:** Raw chicken must be stored below ready-to-eat food, such as pumpkin pie, if it is stored in the same walk-in refrigerator. *(See page 7-7.)*

❺ **True or False:** If stored food has passed its expiration date, you should cook and serve it at once. *(See page 7-4.)*

For answers, please turn to page 7-19.

CONCEPTS

▶ **Refrigerated storage:** Storage used for holding potentially hazardous food at an internal temperature of 41°F (5°C) or lower. (Some jurisdictions allow food in refrigerators to be held at an internal temperature of 45°F [7°C] or lower.) **Check with the local regulatory agency for specific regulations.**

▶ **Frozen storage:** Storage typically designed to hold food at 0°F (–18°C) or lower. Some food requires a different temperature.

▶ **Dry storage:** Storage used to hold dry and canned food at temperatures between 50°F and 70°F (10°C and 21°C) and at a relative humidity of fifty to sixty percent.

▶ **First in, first out (FIFO):** Method of stock rotation in which products are shelved based on their use-by or expiration dates, so oldest products are used first.

The Flow of Food

▶ **Shelf life:** Recommended period of time during which food can be stored and remain suitable for use.

▶ **Hygrometer:** Instrument used to measure relative humidity in storage areas.

▶ **UHT and aseptically packaged food:** Food that has been pasteurized at ultra-high temperatures (UHT) and packaged in containers free of microorganisms (aseptically packaged). Some milk and pudding are shipped this way.

INTRODUCTION

When food is stored improperly and not used in a timely manner, quality and safety suffer. Poor storage practices can cause food to spoil quickly with serious results.

GENERAL STORAGE GUIDELINES

▶ **Label food.** All potentially hazardous, ready-to-eat food, prepared on-site that has been held for longer than twenty-four hours must be labeled with either the date it was prepared, or the date it should be sold, consumed, or discarded. If an item has been previously cooked and stored and is later mixed with another food item to make a new dish, the label on the new dish must indicate the preparation or discard date for the previously cooked item. For example, if ground beef has been cooked and stored at 41°F (5°C) or lower, and later is used to make meat sauce, the meat sauce must be labeled with either the preparation or discard date of the ground beef.

▶ **Rotate products to ensure that the oldest inventory is used first.** The first in, first out (FIFO) method is commonly used to ensure that refrigerated, frozen, and dry products are properly rotated during storage. By this method, a product's use-by, expiration, or preparation date is first identified. The products are then stored to ensure that the oldest are used first. One way to do this is to train employees to store products with the earliest use-by or expiration dates in front of products with later dates. Once shelved, those stored in front are used first. (See *Exhibit 7a.*)

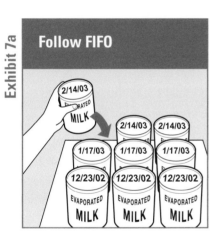

Exhibit 7a

Follow FIFO

Products can be shelved so that those with the earliest use-by, expiration, or preparation dates are stored in front of those with later dates. Once shelved, those stored in front are used first.

Potentially hazardous, ready-to-eat food prepared in-house can be stored for a maximum of seven days at 41°F (5°C) or lower.

▶ **Discard food that has passed its manufacturer's expiration date.** All potentially hazardous, ready-to-eat food that has been prepared in-house can be stored for a maximum of seven days at 41°F (5°C) or lower before it must be discarded.

▶ **Establish a schedule to ensure that stored product is depleted on a regular basis.** If the product has not been sold or consumed by a pre-determined date, discard it, clean and sanitize the container, and refill the container with new product.

▶ **Transfer food between containers properly.** If you take food out of its original package, put it in a clean, sanitized container and cover it. The new container must be labeled with the name of the food being stored and its original use-by or expiration date. Never use empty food containers to store chemicals or put food in empty chemical containers.

▶ **Keep potentially hazardous food out of the temperature danger zone.** Store deliveries as soon as they have been inspected. Take out only as much food as you can prepare at one time. Put prepared food away until needed. Properly cool and store cooked food as soon as it is no longer needed. *(See Section 8 for more information on cooling cooked food.)*

▶ **Check temperatures of stored food and storage areas.** Temperatures should be checked at the beginning of the shift. Many establishments use a preshift checklist to guide employees through this process.

▶ **Store food only in designated storage areas.** Do not store food products near chemicals or cleaning supplies in restrooms, locker rooms, janitor closets, furnace rooms, vestibules, or under stairways or pipes of any kind. Food can easily be contaminated in any of these areas.

▶ **Keep all storage areas clean and dry.** Floors, walls, and shelving in refrigerators, freezers, dry storerooms, and heated holding cabinets should be properly cleaned on a regular basis. Clean up spills and leaks right away to keep them from contaminating other food.

▶ **Clean dollies, carts, transporters, and trays often.**

The Flow of Food

Apply Your Knowledge **To Pitch or Not to Pitch**

Several ready-to-eat items were prepared in-house and stored in this refrigerator at 39°F (4°C). **Today's date is October 3.** Circle food items that should have been discarded. Remember that there are thirty days in September.

For answers, please turn to page 7-19.

REFRIGERATED STORAGE

These areas are typically used to hold potentially hazardous food at 41°F (5°C) or lower. **Some jurisdictions allow food to be held at 45°F (7°C) or lower.** Refrigeration slows the growth of microorganisms and helps keep them from multiplying to levels high enough to cause illness.

Exhibit 7b

Monitor Refrigerated Food Temperature Regularly

Randomly sample the internal temperature using a calibrated thermometer.

The following guidelines should be followed when you are storing food in refrigerators:

▶ **To hold food at a specific internal temperature, refrigerator air temperature should be at least 2°F (1°C) lower than the desired temperature.** For example, to hold poultry at an internal temperature of 41°F (5°C), the air temperature in the refrigerator should be at least 39°F (4°C). At least once during each shift, check the temperature of the unit. Use hanging thermometers in the warmest part of the refrigerator. Some units have a temperature readout on the outside. These should also be checked for accuracy.

▶ **Monitor food temperature regularly.** Randomly sample the temperature of the food stored inside with a calibrated thermometer. (See *Exhibit 7b.*) You may also want to monitor food temperature through the use of a product-mimicking device.

▶ **Do not overload the refrigerator.** Storing too many products prevents good airflow and makes the unit work harder to stay cold.

▶ **Use open shelving.** Lining shelves with aluminum foil or paper restricts circulation of cold air in the unit.

▶ **Never place hot food in the refrigerator.** This can warm the interior enough to put other food in the temperature danger zone.

▶ **Keep the refrigerator door closed as much as possible.** Frequent opening lets warm air inside, which can affect food safety and make the unit work harder. Consider using cold curtains to help maintain walk-in refrigerator temperature.

▶ **Store raw meat, poultry, and fish separately from cooked and ready-to-eat food to prevent cross-contamination.** If they cannot be stored separately, store cooked or ready-to-eat food above raw meat, poultry, and fish. This will prevent raw product juices from dripping onto the prepared food and causing a foodborne illness. It is also recommended that raw meat, poultry, and fish be stored in the following top-to-bottom order in the refrigerator: whole fish, whole cuts of beef and pork, ground meats and fish, whole and ground poultry. This order is based upon the required minimum internal cooking temperature of each food.

▶ **Wrap food properly.** Leaving food uncovered can lead to cross-contamination.

FROZEN STORAGE

Freezing does not kill all microorganisms; however, it slows their growth substantially. When storing food in freezers, you should follow these guidelines.

▶ **Keep freezer temperature at 0°F (–18°C) or lower unless the food you are storing requires a different temperature.**

▶ **Check unit temperatures regularly.**

▶ **Store food at a temperature that will keep it frozen.** This temperature will vary from product to product. A temperature that is good for one product may affect the quality of another.

▶ **Place frozen food deliveries in the freezer as soon as they have been inspected.** Clearly label the food, identifying the package's contents, date of delivery, and use-by date if there is one. Never hold frozen food at room temperature.

▶ **Use caution when placing food into a freezer.** Warm food can raise the temperature inside the unit and partially thaw the food inside. Store food to allow good air circulation. Overloading a freezer makes it work harder, and makes it harder to find and rotate food properly.

Apply Your Knowledge   **Load the Fridge!**

Store each food on the proper storage shelf by writing the letter of the item next to the number on the shelf.

Ⓐ
Whole, raw meat

Ⓑ
Raw poultry

Ⓒ
Cooked and ready-to-eat food

Ⓓ
Whole, raw fish

Ⓔ
Raw, ground meat

For answers, please turn to page 7-19.

▶ **Defrost freezers regularly.** They will operate more efficiently when free of frost. Move food to another freezer while defrosting.

▶ **Keep the unit closed as much as possible.** Use cold curtains to help maintain temperatures.

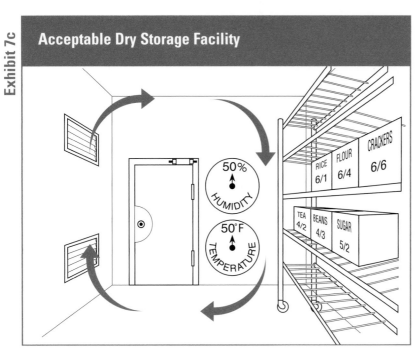

Acceptable Dry Storage Facility

50% HUMIDITY

50°F TEMPERATURE

RICE 6/1 FLOUR 6/4 CRACKERS 6/6

TEA 4/2 BEANS 4/3 SUGAR 5/2

Dry-storage temperatures should be between 50°F and 70°F (10°C and 21°C), and the humidity should be between fifty and sixty percent, if possible.

DRY STORAGE

You should follow these guidelines when placing food in dry storage:

▶ **Keep storerooms cool, dry, and well ventilated.** Moisture and heat are the biggest dangers to dry and canned food. The temperature of the storeroom should be between 50°F and 70°F (10°C and 21°C). Keep relative humidity at fifty to sixty percent, if possible. Use a hygrometer to measure humidity. (See *Exhibit 7c.*)

▶ **Store dry food away from walls and at least six inches off the floor.**

▶ **Keep dry food out of direct sunlight.**

▶ **Keep the area clean.**

▶ **Make sure storerooms are well ventilated.** This will help keep temperature and humidity constant throughout the storage area.

STORING SPECIFIC FOOD

Some food has specific storage requirements. *Exhibit 7d* on the next page outlines some of these requirements.

Exhibit 7d

Recommended Requirements for Storing Specific Food		
Product	**Storage Temperature**	**Other Requirements**
Meat	Store fresh at an internal temperature of 41°F (5°C) or lower.	► Wrap meat in airtight, moisture-proof material or place it in containers.
Poultry	Store fresh at an internal temperature of 41°F (5°C) or lower.	► Store ice-packed poultry as is, in self-draining containers. Change the ice often and sanitize the container regularly.

Fish	Store fresh at an internal temperature of 41°F (5°C) or lower.	► Store ice-packed fish as is, in self-draining containers. Change the ice often and sanitize the container regularly.
		 ► Keep fillets and steaks in original packaging or tightly wrapped in moisture-proof material.
		 ► Store frozen fish in moisture-proof wrapping.
		 ► Fish that will be served raw or partially cooked (with the exception of certain species of tuna) should be frozen by the processor to the following temperatures prior to shipment:
		  ► −4°F (−20°C) or lower for seven days (168 hours) in a storage freezer; or
		  ► −31°F (−35°C) or lower for fifteen hours in a blast freezer.

Shellfish	Store alive at an air temperature of 45°F (7°C) or lower.	► Store alive in the original container.
		 ► Molluscan shellfish (clams, oysters, mussels, scallops) can be stored in a display tank prior to service under one of two conditions:
		  ► The tanks carry a sign stating that the shellfish are for display only.
		  ► You obtain a variance from the local health department.
		 ► Shellstock tags must be kept on file for ninety days from the date the last shellfish was used.

Eggs (shell)	Store fresh at an air temperature of 45°F (7°C) or lower. Maintain constant temperature and humidity.	► Keep eggs in refrigerated storage until used. ► Use all eggs within four to five weeks of the packing date. ► Liquid egg product should be stored according to the manufacturer's recommendations. ► Dried egg product can be stored in a dry, cool storeroom, but should be refrigerated at 41°F (5°C) or lower when reconstituted (mixed with water).

Recommended Requirements for Storing Specific Food *continued*

Product	Storage Temperature	Other Requirements
Dairy Ice Cream and Frozen Yogurt	Store fresh at a temperature of 41°F (5°C) or lower. Store frozen at a temperature of 6°F to 10°F (−14°C to −12°C).	▶ Use the FIFO method of stock rotation. ▶ Discard products if they have passed their use-by or expiration dates.
Fresh Produce	Storage temperatures vary depending upon the product.	▶ Whole, raw produce and raw, cut vegetables delivered packed on ice can be stored that way. The containers must be self-draining, and ice should be changed regularly. ▶ Most produce should not be washed before storage.
MAP, Vacuum Packed, and *Sous Vide* Packaged Food	Store at temperatures recommended by the manufacturer or at 41°F (5°C) or lower.	▶ Discard product if it has passed its expiration or use-by date. ▶ Discard product if the package is torn or slimy, if it contains excessive liquid, or if the product bubbles, indicating the possible growth of *Clostridium botulinum*.
UHT Products Aseptically Packaged	Store product at room temperature.	▶ Once opened, store UHT food that has been aseptically packaged at 41°F (5°C) or lower.
UHT Products *Not* Aseptically Packaged	Store at 41°F (5°C) or lower.	
Canned and Dry Food	Store at a temperature of 50°F to 70°F (10°C to 21°C).	▶ Keep storerooms dry. ▶ If dry food is removed from original packaging, store it in airtight, clearly labeled containers. ▶ Check packages for insect or rodent damage.

Apply Your Knowledge

Find the ten unsafe storage practices in this picture, and describe them in the space provided.

For answers, please turn to page 7-19.

What's Wrong with This Picture?

❶ _____

❷ _____

❸ _____

❹ _____

❺ _____

❻ _____

❼ _____

❽ _____

❾ _____

❿ _____

SUMMARY

When food is stored improperly, quality and safety will suffer. Although different food has different storage needs, some common rules apply. Food should be stored in designated areas and rotated to ensure that the oldest product is used first. It should also be stored in its original packaging. If food must be removed from its original packaging, wrap it in clean, moisture-proof materials or place it in clean and sanitized containers with tight-fitting lids. Make sure all packaging and containers are labeled with the name of the food being stored. All potentially hazardous, ready-to-eat food that is prepared on-site and held for longer than twenty-four hours must be labeled with either the date it was prepared or the date it should be sold, consumed, or discarded. It can be stored for a maximum of seven days at 41°F (5°C) or lower before it must be discarded. Discard all food that has passed its manufacturer's expiration date. Check the temperatures of stored food and the storage area regularly, and keep these areas clean and dry to prevent contamination.

Hold potentially hazardous food in storage areas refrigerated at 41°F (5°C) or lower—temperatures that slow the growth of microorganisms. To hold food at the proper temperature, keep the air temperature of the unit at least 2°F (1°C) lower than the desired food temperature. Never place hot food in the refrigerator, which could raise the temperature inside. Do not line refrigerator shelves, overload the unit, or open the door too often. These practices make the unit work harder to maintain the temperature inside the unit. If possible, store raw meat, poultry, and fish separately from cooked and ready-to-eat food to prevent cross-contamination. If not, store these items below cooked or ready-to-eat food.

While freezing does not kill microorganisms, it slows their growth substantially. Freezers should be kept at 0°F (–18°C) or lower, unless the food being stored requires a different temperature. Unit temperatures should be checked often.

Dry-storage areas should be kept at the appropriate temperature and humidity levels and should be clean and well ventilated to maintain food quality. Food in dry storage should be stored away from walls and at least six inches off the floor. Do not store food products near chemicals or cleaning supplies since food can easily become contaminated. Empty food containers should never be used to store chemicals.

Fresh meat, poultry, fish, and dairy products should be stored at 41°F (5°C) or lower. Fish and poultry can be stored under refrigeration in crushed ice as long as the containers are self-draining, the ice is changed, and the container is sanitized regularly. Eggs should be refrigerated at an ambient (air) temperature of 45°F (7°C) or lower, right up until they are used.

Live, molluscan shellfish should be stored in their original containers at an ambient temperature of 45°F (7°C). Fresh produce has various temperature requirements for storage. Produce should not be washed before storage because it can promote mold growth. MAP, vacuum-packed, and *sous vide* food should be stored at temperatures recommended by the manufacturer. Packages should be checked for signs of contamination, including bubbling, excessive liquid, tears, and slime. Once opened, UHT and aseptically packaged food should be stored in the refrigerator at 41°F (5°C) or lower. Dry and canned food should be stored at temperatures between 50°F and 70°F (10°C to 21°C).

Apply Your Knowledge   A Case in Point 1

❶ What storage errors
were made?

❷ What food items are at risk?

**For answers, please turn
to page 7-19.**

On Monday afternoon, the kitchen staff at the Sunnydale Nursing Home was busy cleaning up after lunch and preparing for dinner. Pete, a kitchen assistant, put a large stockpot of hot, leftover vegetable soup in the refrigerator to cool. Angie, a cook, began deboning the chicken breasts that were stored earlier. When she finished, she put the chicken on an uncovered sheet pan and stored the pan in the refrigerator. She carefully placed the raw chicken on the top shelf, away from the hot soup. Next, Angie iced a carrot cake she had baked that morning. She put the carrot cake in the refrigerator on the shelf directly below the chicken breasts.

Apply Your Knowledge   **A Case in Point 2**

❶ What storage mistakes were made at the restaurant?

For answers, please turn to page 7-19.

A shipment was delivered to Enrico's Italian Restaurant on a warm summer day. Alyce, who was in charge of receiving for the restaurant, inspected the shipment and immediately proceeded to store the items. She loaded a case of sour cream on the dolly and wheeled it over to the reach-in refrigerator. When she opened the refrigerator, she noticed that it was tightly packed, however, she was able to squeeze the case into a spot on the top shelf.

Next, Alyce wheeled several cases of fresh ground beef over to the walk-in refrigerator. She noticed that the readout on the outside of the walk-in indicated 39°F (4°C). Alyce pushed through the cold curtains and bumped into a hot stockpot of soup as she moved inside. She moved the soup over and made a space next to the door for the ground beef. Alyce said hello to Mary, who had just cleaned the shelving in the unit and was lining it with new aluminum foil.

Alyce returned to the receiving area and loaded several cases of pasta on the dolly. She was sweating as she stacked the boxes on the shelving unit and gave a quick glance at the thermometer in the dry-storage room, which read 85°F (29°C). When she was finished stacking the boxes, Alyce returned the dolly to the receiving area.

Apply Your Knowledge

Use these questions to test your knowledge of the concepts presented in this section.

Multiple-Choice Study Questions

1. At what internal temperature would stored ground beef most likely become unsafe to use?
   A. 0°F (–17°C)   C. 41°F (5°C)
   B. 30°F (–1°C)   D. 60°F (16°C)

2. Dry-storage rooms should be kept at
   A. 35°F to 41°F (2°C to 5°C).
   B. 45°F to 50°F (7°C to 10°C).
   C. 50°F to 70°F (10°C to 21°C).
   D. 70°F to 80°F (21°C to 27°C).

3. Under which condition could you use a tank to display live mussels you will be serving to customers?
   A. You have obtained a variance from the health department.
   B. You clean the display tank at least once a month.
   C. You will mix them with other mussels that have not been on display.
   D. You have removed the shellstock identification tags as required by law.

4. Which of the following is *not* a good storage practice?
   A. Shelving food based on its expiration date
   B. Storing raw poultry at temperatures between 41°F and 135°F (5°C and 57°C)
   C. Storing live shellfish at an ambient temperature of 45°F (7°C) or lower
   D. Storing raw meat below ready-to-eat food

5. The first in, first out (FIFO) method helps ensure all of the following during storage *except*
   A. products are properly rotated.
   B. the oldest products are used first.
   C. prepared items are used before they expire.
   D. items that have passed their expiration date are used first.

Continued on next page...

Apply Your Knowledge   **Multiple-Choice Study Questions** *continued*

6. When storing products using the FIFO method, the products with the earliest use-by dates should be
   A. stored in front of products with later use-by dates.
   B. stored behind products with later use-by dates.
   C. stored alongside products with later use-by dates.
   D. stored away from products with later use-by dates.

7. A restaurant that has prepared tuna salad can store it at 41°F (5°C) for a maximum of
   A. 1 day.
   B. 3 days.
   C. 7 days.
   D. 14 days.

8. To keep refrigerated food at an internal temperature of 41°F (5°C), the air temperature in the refrigerator should be at least
   A. 0°F (–18°C).
   B. 26°F (–3°C).
   C. 32°F (0°C).
   D. 39°F (4°C).

9. All of the following are incorrect storage practices *except*
   A. lining refrigerator shelving with aluminum foil.
   B. cooling hot food in a refrigerator.
   C. storing products with the earliest expiration dates in front of products with older dates.
   D. storing fresh poultry over ready-to-eat food.

10. Which storage practice is incorrect?
   A. Storing fresh lamb at 41°F (5°C)
   B. Storing eggs at room temperature
   C. Storing raw clams in their shipping crate at an ambient temperature of 45°F (7°C)
   D. Storing raw ground beef in its original packaging

For answers, please turn to page 7-19.

Apply Your Knowledge   Answers

Page	Activity

7-2   Test Your Food Safety Knowledge

1. False   2. False   3. True   4. True   5. False

7-5   To Pitch or Not To Pitch

Potentially hazardous, ready-to-eat food that has been prepared in-house can be stored for a maximum of seven days at 41°F (5°C) or lower before it should be discarded. The following items should have been discarded: ❶ Tuna salad (stored for eight days) ❷ Vegetable soup (The soup prepared on 9/24 was stored for nine days; The soup prepared on 9/25 was stored for eight days.) **Note:** The salad is not potentially hazardous.

7-8   Load the Fridge!

1. C   2. D   3. A   4. E   5. B

7-12   What's Wrong with This Picture?

❶ The chef is going to place a large amount of hot food into the refrigerator. ❷ The chef is going to store an uncovered stockpot of cooked food in the refrigerator. ❸ The refrigerator is overloaded. ❹ The chef is smoking a cigarette. ❺ The chef has an uncovered bandage on his hand. ❻ Some items in the refrigerator are labeled while others are not. ❼ Food has been stored on the floor, and one of the bags must have a tear as it appears food has spilled onto the floor. ❽ Chemicals and cleaning tools have been stored with food. ❾ Many of the dry goods stored on the shelf have not been labeled. ❿ From the cobwebs, it appears the establishment does not do a good job cleaning this storeroom or rotating stock.

7-15   A Case in Point 1

❶ Pete and Angie made several errors. Ⓐ Pete should not have placed the hot stockpot of soup into the refrigerator to cool. Not only is this an unsafe way to cool hot, potentially hazardous food, but the temperature of the pot could warm the refrigerator interior enough to put other stored food in the TDZ. Pete should have cooled the soup properly *(see Section 8)* before storing it in the refrigerator. Ⓑ Angie should not have placed the uncovered pan of raw chicken on the top shelf in the refrigerator, nor should she have stored the carrot cake below the raw chicken breasts, since the juices could have dripped onto the cake, contaminating it. Cooked or ready-to-eat food must be stored above raw meat, poultry, and fish, if these items are stored in the same unit. Ideally, raw product like fresh meat, fish, and poultry should be stored in a separate unit from cooked and ready-to-eat food. ❷ Unfortunately, the mistakes they made could affect all of the food stored in the refrigerator.

7-16   A Case in Point 2

❶ The following mistakes were made at the restaurant: Ⓐ Alyce placed the case of sour cream into an already overloaded refrigerator. Ⓑ A hot stockpot of soup was stored in the walk-in refrigerator. Hot food should never be placed in a refrigerator. Ⓒ Mary was lining the refrigerator shelving with aluminum foil. This can restrict airflow in the unit. Ⓓ The temperature in the dry storage room was 85°F (29°C), which is too warm. Dry-storage areas should be between 50°F and 70°F (10°C and 21°C).

7-17   Multiple-Choice Study Questions

1. D   3. A   5. D   7. C   9. C
2. C   4. B   6. A   8. D   10. B

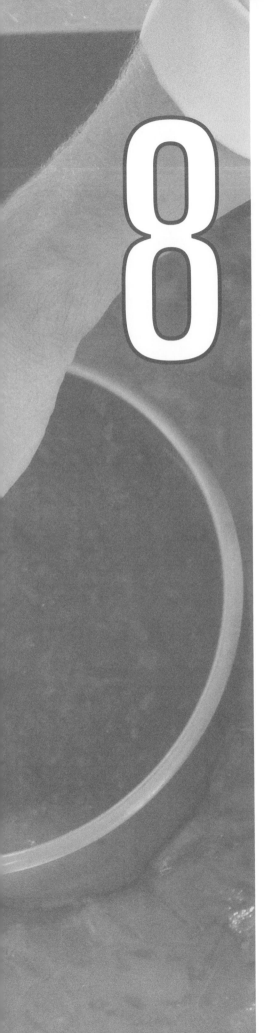

The Flow of Food: Preparation

Inside this section:

▶ Thawing Food Properly
▶ Preparing Specific Food
▶ Cooking Food

▶ Cooling Food
▶ Storing Cooked Food
▶ Reheating Potentially Hazardous Food

After completing this section, you should be able to:

▶ Identify proper methods for thawing food.
▶ Identify the minimum internal cooking times and temperatures for potentially hazardous food.
▶ Identify the proper procedure for cooking potentially hazardous food in a microwave.
▶ Identify methods and time and temperature requirements for cooling cooked food.

▶ Identify time and temperature requirements for reheating cooked, potentially hazardous food.
▶ Identify methods for preventing contamination and time and temperature abuse when preparing food.
▶ Recognize the importance of informing consumers of risks when serving raw or undercooked food.

Apply Your Knowledge

Check to see how much you know about the concepts in this section. Use the page references provided to explore the topic in each question.

Test Your Food Safety Knowledge

❶ **True or False:** Ground beef should be cooked to a minimum internal temperature of 140°F (60°C). *(See page 8-10.)*

❷ **True or False:** Fish cooked in a microwave must be heated to 145°F (63°C). *(See page 8-11.)*

❸ **True or False:** Cooked, potentially hazardous food must be cooled from 135°F to 70°F (57°C to 21°C) within four hours and from 70°F to 41°F (21°C to 5°C) or lower within an additional two hours. *(See page 8-14.)*

❹ **True or False:** When potentially hazardous food is reheated for hot-holding, it must be heated to 155°F (68°C) for fifteen seconds within two hours. *(See page 8-17.)*

❺ **True or False:** It is acceptable to thaw a beef roast at room temperature. *(See page 8-3.)*

For answers, please turn to page 8-22.

CONCEPTS

▶ **Minimum internal cooking temperature:** The required cooking temperature the internal portion of food must reach in order to sufficiently reduce the number of microorganisms that might be present. This temperature is specific to the type of food being cooked. Food must reach and hold its minimum internal temperature for a specified amount of time.

▶ **Two-stage cooling:** Criteria by which cooked food must be cooled from 135°F to 70°F (57°C to 21°C) within two hours and from 70°F (21°C) to 41°F (5°C) or lower within an additional four hours, for a total cooling time of six hours.

▶ **Ice-water bath:** Method of cooling food in which a container holding hot food is placed into a larger container of ice water.

▶ **Cold paddle:** Plastic paddle filled with water and frozen. When used to stir hot food, it cools the food quickly.

INTRODUCTION

Once food has been received and stored safely, it is essential that it be prepared, cooked, cooled, and reheated with just as much care. It is at these points in the flow of food that the risk of cross-contamination and time-temperature abuse are the greatest.

THAWING FOOD PROPERLY

Freezing does not kill microorganisms, but it does slow their growth. When frozen food is thawed and exposed to the temperature danger zone, any foodborne microorganisms present will begin to grow and multiply. There are only four acceptable methods for thawing potentially hazardous frozen food. (See *Exhibit 8a*.)

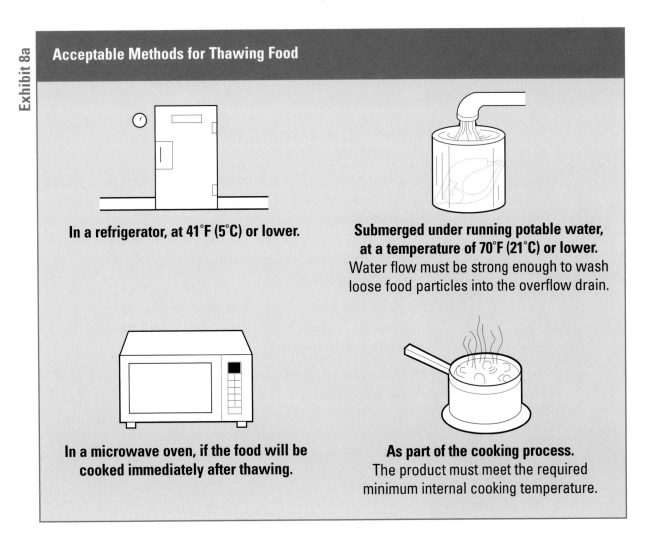

Exhibit 8a

Acceptable Methods for Thawing Food

In a refrigerator, at 41°F (5°C) or lower.

Submerged under running potable water, at a temperature of 70°F (21°C) or lower. Water flow must be strong enough to wash loose food particles into the overflow drain.

In a microwave oven, if the food will be cooked immediately after thawing.

As part of the cooking process. The product must meet the required minimum internal cooking temperature.

PREPARING SPECIFIC FOOD

Practicing time and temperature control and taking steps to prevent cross-contamination will help you avoid most cases of foodborne illness. However, some food and types of preparation require a bit more care.

Meat, Fish, and Poultry

Follow these additional guidelines when working with raw meat, poultry, and seafood:

► **Use clean and sanitized work areas, cutting boards, knives, and utensils.**

► **Wash hands properly.** If gloves are worn, change them before starting each new task.

► **Remove from refrigerated storage only as much product as you can prepare at one time.**

► **Return raw prepared meats to refrigeration, or cook them as quickly as possible.** Store properly to prevent cross-contamination.

Salads Containing Potentially Hazardous Food

Chicken, tuna, egg, pasta, and potato salads have been known to cause foodborne-illness outbreaks. Since they are not typically cooked after preparation, there is no chance to kill microorganisms that may have been introduced during preparation. Therefore, care must be taken when preparing these salads. Follow these preparation guidelines:

► **Make sure leftover meat and poultry have been properly cooked, held, cooled, and stored before adding them to salad.**

► **Make sure leftovers used for salad have not been left in the refrigerator too long.** Leftover meat and poultry should be discarded after seven days if stored under refrigeration at 41°F (5°C) or lower (four days if stored at 45°F [7°C]). Salad made with these items should be discarded when this storage period expires.

▶ **Leave food in the refrigerator until all ingredients are ready to be mixed.**

▶ **Consider chilling all ingredients and utensils before using them to make salad.** For example, tuna, mayonnaise, and mixing bowls can be chilled before making tuna salad.

▶ **Prepare food in small batches, so large amounts of food do not sit out at room temperature for long periods of time.**

Eggs and Egg Mixtures

All untreated shell eggs are considered potentially hazardous food because they are able to support the rapid growth of microorganisms. When preparing eggs and egg mixtures, follow these guidelines:

▶ **Handle pooled eggs (if allowed) with special care.**
Pooled eggs are eggs that are cracked open and combined in a common container. They must be handled with care because bacteria in one egg can be spread to the rest. Pooled eggs must be cooked promptly after mixing, or stored at 41°F (5°C) or lower. Containers that have been used to hold pooled eggs must be washed and sanitized before being used for a new batch.

▶ **Operations that serve high-risk populations, such as those in hospitals and nursing homes, must always use pasteurized eggs or egg products.**

▶ **Promptly clean and sanitize all equipment and utensils used to prepare eggs.**

▶ **Consider using pasteurized shell eggs or pasteurized egg products when preparing egg dishes requiring little or no cooking.** These include dishes such as mayonnaise, eggnog, Caesar salad dressing, and hollandaise sauce.

Exhibit 8b

Breading Food

Never use breading for more than one product.

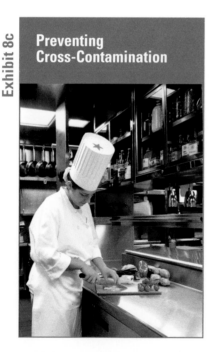

Exhibit 8c

Preventing Cross-Contamination

Make sure fruit and vegetables do not come in contact with surfaces exposed to raw meat and poultry.

Reprinted with permission from Tony Soluri and Charlie Trotter.

Batter and Breading

If prepared with milk or eggs, batter and breading are at risk for time-temperature abuse and cross-contamination, so they should be handled with care. If you make breaded or battered food from scratch, follow these guidelines:

▶ **Prepare batter in small batches.** Store what you do not need at 41°F (5°C) or lower in a covered container.

▶ **When breading food that will be cooked at a later time, store it in the refrigerator as quickly as possible.**

▶ **Throw out any unused batter or breading after each shift.** Never use batter or breading for more than one product. (See *Exhibit 8b*.)

▶ **Consider making batter with pasteurized shell eggs or egg products whenever possible.**

Fruit and Vegetables

When preparing fruit and vegetables, follow these guidelines:

▶ **Make sure fruit and vegetables do not come in contact with surfaces exposed to raw meat and poultry.** Prepare fruit and vegetables away from raw meat, poultry, eggs, and cooked and ready-to-eat food. (See *Exhibit 8c*.) Clean and sanitize the work space and all utensils that will be used during preparation.

▶ **Wash fruit and vegetables thoroughly under running water to remove dirt and other contaminants before cutting, cooking, or combining with other ingredients.** Pay particular attention to leafy greens, such as lettuce and spinach. Remove the outer leaves, and pull lettuce and spinach completely apart and rinse thoroughly.

▶ **Refrigerate and hold cut melons at 41°F (5°C) or lower since they are potentially hazardous food.**

▶ **Do not add sulfites (preservatives that maintain freshness) to food.**

▶ **If your establishment primarily serves high-risk populations, do not serve raw seed sprouts.**

Fresh Juice

Many establishments prepare fresh fruit and vegetable juice on-site for their customers. However, if the juice is packaged on-site for sale at a later time, the establishment must have a variance from the regulatory agency, and the juice must be treated (i.e., pasteurized) according to an approved HACCP plan or be labeled with the following phrase: *"Warning: This product has not been pasteurized and therefore may contain harmful bacteria that can cause serious illness in children, the elderly, and people with weakened immune systems."* Federal public health officials require that establishments serving high-risk populations only serve fresh juice that has been treated to eliminate pathogens that can cause foodborne illness.

Ice

▶ **Ice used as food or used to chill food must be made from drinking water.**

▶ **Ice used to chill food or beverages should never be used as an ingredient.**

▶ **Use a clean, sanitized container and ice scoop to transfer ice from an ice machine to other containers.** Never hold or transport ice in containers that have held raw meat, fish, poultry, or chemicals. Store ice scoops outside of the ice machine in a sanitary, protected location. (See *Exhibit 8d.*) Never use a glass to scoop ice, and never let your hands come in contact with it.

Preparation Practices That Require a Variance

Some preparation methods require a variance from the regulatory agency. A variance is required whenever an establishment

▶ smokes food or uses food additives as a method of food preservation.

▶ cures food.

▶ custom-processes animals for personal use.

▶ packages food using a reduced-oxygen packaging method (MAP, vacuum-packaging).

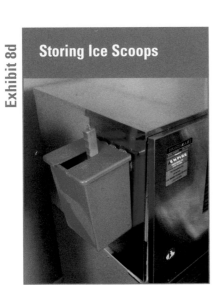

Exhibit 8d

Storing Ice Scoops

Store ice scoops outside of the ice machine in a sanitary, protected container.

COOKING FOOD

Handling food safely prior to cooking is very important. While cooking food to the required minimum internal temperature is the only way to sufficiently reduce the number of foodborne microorganisms to safe levels, it does not destroy spores or toxins these microorganisms create. Safe handling of the product before it is cooked is essential to preventing microorganisms from growing and producing spores and toxins.

The minimum internal temperature at which foodborne microorganisms are destroyed varies from product to product. Minimum standards have been developed for most food. (See *Exhibit 8e.*) These temperatures must be reached and held for the specified amount of time. Use a properly calibrated thermometer with a suitable size probe to measure the internal temperature of food. Measure internal temperature in the thickest part of the food, and take at least two readings in different locations.

It is important to remember that potentially hazardous food— such as meat, eggs, seafood, and poultry—should be cooked to the minimum internal temperatures specified in this section unless otherwise ordered by the customer. Potentially hazardous food not cooked to these temperatures—including over-easy eggs, raw oysters, sashimi, and rare hamburgers—generally do not pose an unacceptable risk of foodborne illness to the healthy customer. However, if a customer is in a group at risk for foodborne illness *(see Section 1),* consuming raw or undercooked, potentially hazardous food could possibly increase the risk of illness, sometimes seriously.

In all cases, high-risk customers should be advised of the potential risk if they ask about or specifically request undercooked food or any potentially hazardous food (or ingredient) that is raw or not fully cooked. They may want to consult with a physician before regularly consuming these types of food. **Additionally, check with your regulatory agency for specific requirements.**

Cooking Requirements for Specific Types of Food

Product	Minimum Internal Cooking Temperature	Other Cooking Requirements and Recommendations
Poultry (including whole or ground duck, chicken, turkey)	**165°F (74°C)** for 15 seconds	Poultry has more types and higher counts of microorganisms than other meat. Therefore, it should be cooked more thoroughly.
Stuffing and Stuffed Meat	**165°F (74°C)** for 15 seconds	Stuffing poses a hazard because ▶ it can be made with potentially hazardous food. ▶ it acts as insulation, preventing heat from reaching the center of meat or poultry. Stuffing should be cooked separately, particularly when cooking whole, large birds or large cuts of meat.
Dishes That Include Potentially Hazardous Ingredients ▶ When including *previously cooked,* potentially hazardous ingredients in the dish	Cook the ingredients to **165°F (74°C)** for 15 seconds.	Example: when adding cooked ground beef to a red sauce, the beef within the dish must be reheated to 165°F (74°C) for 15 seconds within 2 hours.
▶ When including *raw* potentially hazardous ingredients in the dish	Cook the raw ingredients to their required minimum internal temperature.	Example: when cooking jambalaya you must ensure that the raw shrimp reaches 145°F (63°C) for 15 seconds.

Cooking Requirements for Specific Types of Food *continued*

Product	Minimum Internal Cooking Temperature	Other Cooking Requirements and Recommendations
Ground Meats (including beef, pork, and other meat or fish)	**155°F (68°C)** for 15 seconds	Most whole-muscle cuts of meat are likely to have microorganisms only on their surface. When meat is ground, microorganisms on the surface are mixed throughout the product. Ground meat may also be cooked to the following alternative internal temperatures: ▶ 145°F (63°C) for 3 minutes ▶ 150°F (66°C) for 1 minute ▶ 155°F (68°C) for 15 seconds ▶ 158°F (70°C) <1 second
Injected Meats (including brined ham and flavor-injected roasts)	**155°F (68°C)** for 15 seconds	When meats are injected, foodborne microorganisms on the surface can be carried into the interior. Injected meat may also be cooked to the following alternative internal temperatures: ▶ 145°F (63°C) for 3 minutes ▶ 150°F (66°C) for 1 minute ▶ 155°F (68°C) for 15 seconds ▶ 158°F (70°C) <1 second

Cooking Requirements for Specific Types of Food *continued*

Product	Minimum Internal Cooking Temperature	Other Cooking Requirements and Recommendations
Pork, Beef, Veal, Lamb	Steaks/Chops: **145°F (63°C)** for 15 seconds	This temperature is high enough to destroy *Trichinella* spp. larvae that might have contaminated pork.
	Roasts: **145°F (63°C)** for 4 minutes	Depending upon the type of roast and the oven used, roasts may be cooked to the following alternative internal temperatures: ▶ 130°F (54°C) for 112 minutes ▶ 131°F (55°C) for 89 minutes ▶ 133°F (56°C) for 56 minutes ▶ 135°F (57°C) for 36 minutes ▶ 136°F (58°C) for 28 minutes ▶ 138°F (59°C) for 18 minutes ▶ 140°F (60°C) for 12 minutes ▶ 142°F (61°C) for 8 minutes ▶ 144°F (62°C) for 5 minutes ▶ 145°F (63°C) for 4 minutes
Fish	**145°F (63°C)** for 15 seconds	
Stuffed Fish (or stuffing containing fish)	**165°F (74°C)** for 15 seconds	
Ground, Chopped, or Minced Fish	**155°F (68°C)** for 15 seconds	

Cooking Requirements for Specific Types of Food *continued*		
Product	**Minimum Internal Cooking Temperature**	**Other Cooking Requirements and Recommendations**
Shell Eggs for Immediate Service	**145°F (63°C)** for 15 seconds	When cooking eggs, remove from storage only as many eggs as you need for immediate use. Never stack egg trays (flats) near the grill or stove.
Shell Eggs That Will Be Hot-Held for Service	**155°F (68°C)** for 15 seconds	
Fruit or Vegetables That Will Be Hot-Held for Service	**135°F (57°C)**	Cooked vegetables must never be held at room temperature.
Commercially Processed, Ready-to-Eat Food That Will Be Hot-Held for Service	**135°F (57°C)** for 15 seconds	This includes items such as cheese sticks, deep-fried vegetables, chicken wings, etc.
Potentially Hazardous Food Cooked in the Microwave (eggs, poultry, fish, and meat)	**165°F (74°C)**	► Cover it to prevent the surface from drying out. ► Rotate or stir it halfway through the cooking process to distribute the heat more evenly. ► Let it stand for at least 2 minutes after cooking to let the product temperature equalize. ► Check the temperature in several places to ensure that it is cooked through.

Apply Your Knowledge

Identify the required minimum internal cooking temperature for each food and write the letter in the space provided. Some letters will be used more than once.

What's the Temperature?

___ **1** Ahi tuna steak

___ **2** Green beans that will be hot-held

___ **3** Ground pork

___ **4** Lamb chops

___ **5** Shell eggs for immediate service

___ **6** Duck

___ **7** Precooked frozen hot wings

___ **8** Steak

___ **9** Chicken enchiladas prepared with previously cooked chicken

___ **10** Pork loin injected with marinade

    A. 135°F (57°C)

    B. 145°F (63°C)

    C. 155°F (68°C)

    D. 165°F (74°C)

For answers, please turn to page 8-22.

Apply Your Knowledge

For each statement about microwave cooking, write a **T** in the space provided if the statement is true and an **F** in the space if it is false.

What's Your Microwave IQ?

___ **1** Leave food uncovered so that excess moisture can escape.

___ **2** Cooked food should be left to stand for thirty seconds to allow the temperature to equalize.

___ **3** Fish cooked in a microwave should be heated to 145°F (63°C).

___ **4** Food should be stirred halfway through the cooking process.

___ **5** Food should be rotated right before it is removed from the microwave.

For answers, please turn to page 8-22.

Exhibit 8f

Two-Stage Cooling

135°F
57°C

70°F
21°C

41°F
5°C

Food must be cooled from 135°F to 70°F (57°C to 21°C) within two hours and from 70°F to 41°F (21°C to 5°C) or lower in an additional four hours.

COOLING FOOD

You have already seen how important it is to keep food out of the temperature danger zone. When cooked food will not be served immediately, it is essential to hold it properly or to cool it as quickly as possible.

The FDA Food Code recommends two-stage cooling. Cooked food must be cooled from 135°F (57°C) to 70°F (21°C) within two hours and from 70°F (21°C) to 41°F (5°C) or lower in an additional four hours, for a total cooling time of six hours. (See *Exhibit 8f*.) **Some jurisdictions use one-stage cooling, by which food must be cooled to 41°F [5°C] or lower in less than four hours.**

Keep in mind this is a two-stage process (two hours plus four hours). While it is true that microorganisms grow well in the temperature danger zone, the temperature range between 125°F (52°C) and 70°F (21°C) is ideal for the growth of pathogenic microorganisms. Food must pass through this temperature range quickly during cooling to minimize their growth. Because only two hours are allotted to cool food from 135°F (57°C) to 70°F (21°C), the two-stage cooling process passes potentially hazardous food through this temperature range quickly—and safely.

When using two-stage cooling, if the food has not reached 70°F (21°C) within two hours, it must be discarded or properly reheated. Reheat the food to 165°F (74°C) for fifteen seconds within two hours and then cool it properly.

Methods for Cooling Food

Several factors affect how quickly food will cool. These include:

▶ **The thickness or density of the food being cooled.** The denser the food, the more slowly it will cool. For example, refried beans will take longer to cool than vegetable broth since the beans are denser.

▶ **The container in which a food item is stored.** Stainless steel transfers heat from food faster than plastic. Shallow pans allow the heat from food to disperse faster than deep pans.

Refrigerators are designed to keep cold food cold. Most are not designed to cool hot food quickly. Also, placing hot food in a

The Flow of Food

refrigerator or freezer to cool it may not move the food through the temperature danger zone quickly enough.

There are a number of methods that can be used for cooling food quickly and safely. (See *Exhibit 8g.*)

Exhibit 8g

Safe Methods for Cooling Food

Reduce the quantity or size of the food you are cooling. Cut large food items into smaller pieces or divide large containers of food into smaller containers or shallow pans.

Use ice-water baths. After dividing food into smaller quantities, put the pots or pans into a sink or large pot filled with ice water.

Use a blast chiller to cool food before placing it into refrigeration.

Stir food to cool it faster and more evenly. Some manufacturers make plastic paddles that can be filled with water and frozen. Stirring food with these cold paddles cools food quickly.

You can also cool food using one of the following methods:

▶ **Add ice or cool water as an ingredient.** This method works for recipes requiring water as an ingredient, such as soup or stew. The recipe initially can be prepared with less water than is required. Cold water or ice can then be added after cooking to cool the product and to provide the remaining water required by the recipe.

▶ **If properly equipped, steam-jacketed kettles can be used to cool food.** Simply run cold water through the jacket to cool the food in the kettle.

Apply Your Knowledge	Is It Cool Enough?

❶ Is the soup safe to serve?

❷ Why or why not?

For answers, please turn to page 8-22.

> Bill placed a stockpot of soup that had been held at 135°F (57°C) into an ice-water bath to cool at 8:00 A.M. At 10:00 A.M., he checked the temperature and found that it was 90°F (32°C). Bill continued to cool the soup in the ice-water bath, stirring it occasionally. At 11:00 A.M., when the soup had reached 70°F (21°C), he poured it into shallow pans, and placed it on the top shelf in the walk-in cooler.

STORING COOKED FOOD

Once food has cooled to at least 70°F (21°C), it can be stored on the top shelves in the refrigerator. Pans of food should be positioned so air can circulate around them. Be sure employees monitor the temperature of the food to ensure that it cools to 41°F (5°C) or lower in four hours. Follow FIFO when storing food.

REHEATING POTENTIALLY HAZARDOUS FOOD

Previously cooked, potentially hazardous food reheated for hot-holding, must be taken through the temperature danger zone as quickly as possible. Reheat it to an internal temperature of 165°F (74°C) for fifteen seconds within two hours. If the food has not reached this temperature within two hours, discard it.

Food that is reheated for immediate service to a customer, such as the beef in a roast beef sandwich, may be served at any temperature, as long as the food was properly cooked first.

Apply Your Knowledge   **A Case in Point 1**

---

❶ What did John do wrong?

For answers, please turn to page 8-22.

On Friday, John went to work at The Fish House knowing he had a lot to do. After changing clothes and punching in, he took a case of frozen raw shrimp out of the freezer. To thaw it quickly, he put the frozen shrimp into the prep sink and turned on the hot water. While waiting for the shrimp to thaw, John took several fresh, whole fish out of the walk-in refrigerator. He brought them back to the prep area and began to clean and filet them. When he finished, he put the fillets in a pan and returned them to the walk-in refrigerator. He rinsed off the boning knife and cutting board in the sink, and wiped off the worktable with a dish towel.

Next, John transferred the shrimp from the sink to the worktable using a large colander. On the cutting board, he peeled, deveined, and butterflied the shrimp using the boning knife. He put the prepared shrimp in a covered container in the refrigerator, then started preparing fresh produce.

Apply Your Knowledge   **A Case in Point 2**

---

❶ What did Angie do wrong?

For answers, please turn to page 8-22.

By 7:30 P.M., all the residents at Sunnydale Nursing Home had eaten dinner. As she began cleaning up, Angie realized she had a lot of chicken breasts left over. Betty, the new assistant manager, had forgotten to inform Angie that several residents were going to a local festival and would miss dinner.

"No problem," Angie thought. "We can use the leftover chicken to make chicken salad."

Angie left the chicken breasts in a pan on the prep table while she started putting other food away and cleaning up the kitchen. At 9:45 P.M., when everything else was clean, she put her hand over the pan of chicken breasts and decided they were cool enough to handle. She covered the pan with plastic wrap, and put it in the refrigerator.

Three days later, she came in to work on the early shift. Angie decided to make chicken salad from the leftover chicken breasts. After she hung up her coat and put on her apron, Angie took all the ingredients she needed for chicken salad out of the refrigerator and put them on a worktable. Then she started breakfast.

First, she cracked three dozen eggs into a large bowl, added some milk, and set the bowl near the stove. Then she took bacon out of the refrigerator and put it on the worktable next to the chicken salad ingredients. She peeled off strips of bacon onto a sheet pan and put the pan into the oven. After wiping her hands on her apron, she went back to the stove to whisk the eggs and pour them onto the griddle. When they were almost done, Angie scooped the scrambled eggs into a hotel pan and put it in the steam table.

As soon as breakfast was cooked, Angie went back to the prep table to wash and chop celery and cut up the chicken for the salad.

SUMMARY

To protect food during preparation, you must handle it safely. The keys are time and temperature control and the prevention of cross-contamination.

Thaw frozen food in the refrigerator, under cool running water, in a microwave oven, or as part of the cooking process. Never thaw food at room temperature. Have employees prepare food in small batches, use chilled utensils and bowls, and record product temperatures and preparation times.

Cooking can reduce the number of microorganisms in food to safe levels. To ensure that microorganisms are destroyed, food must be cooked to required minimum internal temperatures for a specific amount of time. These temperatures vary from product to product. Cooking does not kill the spores or toxins some microorganisms produce. That is why it is so important to inspect product once it arrives and handle it safely during preparation.

Once food is cooked, it should be served as quickly as possible. If it is going to be stored and served later, it must be cooled rapidly. Cooked food must be cooled from 135°F (57°C) to 70°F (21°C) within two hours and from 70°F (21°C) to 41°F (5°C) or lower in an additional four hours (for a total cooling time of six hours), unless otherwise required by your local health code. Placing large containers of hot food into the refrigerator can put all other stored food in danger. Methods for cooling large quantities of cooked food quickly include dividing it into smaller portions, putting it in shallow stainless steel pans, using an ice-water bath or blast chiller, or stirring it often with cold paddles. When the food is cold enough, store it properly in the refrigerator.

Previously cooked, potentially hazardous food that will be hot-held must be reheated to an internal temperature of 165°F (74°C) for fifteen seconds within two hours before it can be served.

Apply Your Knowledge   **Multiple-Choice Study Questions**

Use these questions to test your knowledge of the concepts presented in this section.

1. Beef stew must be cooled from 135°F (57°C) to 70°F (21°C) within ____ hours and from 70°F (21°C) to 41°F (5°C) or lower in an additional ____ hours.
   A. 4, 2   B. 2, 4   C. 3, 2   D. 2, 3

2. Which of the following is *not* a safe method for thawing frozen food?
   A. Thawing it by submerging it under running water at 70°F (21°C) or lower
   B. Thawing it in the microwave and cooking it immediately afterward
   C. Thawing it at room temperature
   D. Thawing it in the refrigerator overnight

3. Stuffed pork chops must be cooked to a minimum internal temperature of
   A. 135°F (57°C) for fifteen seconds.
   B. 145°F (63°C) for fifteen seconds.
   C. 155°F (68°C) for fifteen seconds.
   D. 165°F (74°C) for fifteen seconds.

4. When reheating potentially hazardous food for hot-holding, reheat the food to
   A. 135°F (57°C) for fifteen seconds within two hours.
   B. 145°F (63°C) for fifteen seconds within two hours.
   C. 155°F (68°C) for fifteen seconds within two hours.
   D. 165°F (74°C) for fifteen seconds within two hours.

5. Meat, poultry, and fish cooked in a microwave must be heated to at least
   A. 140°F (60°C).   C. 155°F (68°C).
   B. 145°F (63°C).   D. 165°F (74°C).

6. Shell eggs that will be cooked and held for later service must be cooked to an internal temperature of
   A. 140°F (60°C) for fifteen seconds.
   B. 145°F (63°C) for fifteen seconds.
   C. 155°F (68°C) for fifteen seconds.
   D. 165°F (74°C) for fifteen seconds.

Continued on next page...

7. All of the following practices can help prevent cross-contamination during food preparation *except*

A. preparing food in small batches.

B. throwing out unused batter or breading after each shift.

C. preparing raw meat at a different time than fresh produce.

D. cleaning and sanitizing pooled egg containers between batches.

8. What is the proper way to cool a large stockpot of clam chowder?

A. Allow the stockpot to cool at room temperature.

B. Put the hot stockpot into the walk-in refrigerator to cool.

C. Divide the clam chowder into smaller containers and place them in an ice-water bath.

D. Put the hot stockpot into the walk-in freezer to cool.

9. All of the following practices can help prevent time and temperature abuse *except*

A. thawing food in a refrigerator at 41°F (5°C).

B. chilling all ingredients used to make tuna salad.

C. leaving food in the refrigerator until all ingredients are ready to be mixed.

D. thawing steaks in a microwave and promptly refrigerating them for later use.

10. Which of the following food has been safely cooked?

A. A hamburger cooked to an internal temperature of 135°F (57°C) for fifteen seconds

B. A pork chop cooked to an internal temperature of 145°F (63°C) for fifteen seconds

C. A whole turkey cooked to an internal temperature of 155°F (68°C) for fifteen seconds

D. A fish fillet cooked to an internal temperature of 135°F (57°C) for fifteen seconds

For answers, please turn to page 8-22.

Apply Your Knowledge   Answers

Page	Activity				

8-2   Test Your Food Safety Knowledge

1. False	2. False	3. False	4. False	5. False

8-13   What's the Temperature?

1. B	3. C	5. B	7. A	9. D
2. A	4. B	6. D	8. B	10. C

8-13   What's Your Microwave IQ?

1. F	2. F	3. F	4. T	5. F

8-16   Is It Cool Enough?

❶ The soup is not safe to serve. ❷ According to the two-stage cooling method, Bill had two hours to get the hot soup from 135°F (57°C) to 70°F (21°C). It took the soup three hours to reach this temperature. The soup should have been properly reheated or discarded.

8-17   A Case in Point 1

❶ John made several errors. Ⓐ He failed to wash his hands before starting work. Ⓑ He failed to thaw the shrimp properly. The shrimp should have been thawed in the refrigerator at 41°F (5°C) or lower. This, however, would have required advance planning. John also could have thawed the shrimp by submerging it under running potable water at a temperature of 70°F (21°C) or lower. Ⓒ He took out more whole fish from the walk-in refrigerator than he could prepare in a short period of time, unnecessarily subjecting the fish to time-temperature abuse. Ⓓ He failed to clean and sanitize the boning knife, cutting board, and worktable properly after cleaning and fileting the fish. Microorganisms that may have been present on the fish could have been transferred to the shrimp that John prepared with the contaminated knife and cutting board. Ⓔ He did not wash his hands after handling the fish and before preparing the fresh produce.

Continued on next page…

Apply Your Knowledge   Answers *continued*

8-18   A Case in Point 2

❶ Angie made several errors. Ⓐ She cooled the leftover chicken breasts improperly. Food should never be left out to cool at room temperature. Angie could have divided the chicken into smaller portions and refrigerated them, or used a blast chiller. Ⓑ She unnecessarily subjected the chicken salad ingredients to time-temperature abuse by leaving them out on the counter while she performed other duties. Angie should have left the ingredients in the refrigerator until she was ready to prepare the salad. Ⓒ She used raw shell eggs in a nursing home where she serves a population at high risk for foodborne illness. She should not have done this, especially since she was cooking the eggs to hold them for later service. Angie should be using pasteurized shell eggs or egg products. Ⓓ She failed to handle the large number of pooled eggs properly. She left a bowl of eggs near a warm stove in the TDZ. She should have pooled a smaller number of eggs and kept them in an ice-bath away from the stove. Also, she failed to check the temperature of the eggs prior to placing them on the steam table. Eggs that will be cooked and held for later service must be cooked to at least 155°F (68°C) for fifteen seconds. Ⓔ She failed to wash her hands before starting to prepare food and did not wash her hands between foodhandling tasks. If there were any microorganisms on her hands, she would have transferred them to the food and food-contact surfaces she handled. Also, Angie wiped her hands on her apron. Wiping hands is not an adequate substitute for proper handwashing.

8-20   Multiple-Choice Study Questions

1. B	3. D	5. D	7. A	9. D
2. C	4. D	6. C	8. C	10. B

The Flow of Food: Service

Inside this section:
▶ General Rules for Holding Food
▶ Serving Food Safely
▶ Off-Site Service

After completing this section, you should be able to:

▶ Identify time and temperature requirements for holding hot and cold, potentially hazardous food.

▶ Identify procedures for preventing time-temperature abuse and cross-contamination when displaying and serving food.

▶ Identify the requirements for using time rather than temperature as the only method of control when holding ready-to-eat food.

▶ Implement methods for minimizing bare-hand contact with ready-to-eat food.

▶ Identify hazards associated with the transportation of food and methods for preventing them.

▶ Identify hazards associated with the service of food off-site and methods for preventing them.

▶ Identify hazards associated with vending food and methods for preventing them.

▶ Prevent customers from contaminating self-service areas.

▶ Prevent employees from contaminating food.

Apply Your Knowledge

Check to see how much you know about the concepts in this section. Use the page references provided to explore the topic in each question.

Test Your Food Safety Knowledge

① **True or False:** Cold, potentially hazardous food must be held at an internal temperature of 41°F (5°C) or lower. *(See page 9-4.)*

② **True or False:** Hot, potentially hazardous food must be held at an internal temperature of 120°F (49°C) or higher. *(See page 9-3.)*

③ **True or False:** Chicken salad can be held at room temperature, if it has a label that specifies it must be discarded after six hours. *(See page 9-4.)*

④ **True or False:** When holding potentially hazardous food for service, the internal temperature must be checked at least every four hours. *(See page 9-3.)*

⑤ **True or False:** Servers can contaminate food simply by handling the food-contact surface of a plate. *(See page 9-6.)*

For answers, please turn to page 9-17.

CONCEPTS

▶ **Hot-holding equipment:** Equipment such as chafing dishes, steam tables, and heated cabinets specifically designed to hold food at 135°F (57°C) or higher.

▶ **Cold-holding equipment:** Equipment specifically designed to keep cold food at 41°F (5°C) or lower.

▶ **Sneeze guard:** Food shield used over self-service displays and food bars, which extends seven inches beyond the food and fourteen inches above the food counter.

▶ **Off-site service:** Service of food to someplace other than where it is prepared or cooked, including catering and vending.

▶ **Vending machine:** Machine that dispenses hot and cold food, beverages, and snacks.

▶ **Single-use item:** Disposable tableware or packaged food designed to be used only once, including plastic flatware, paper or plastic cups, plates and bowls, as well as single-serve food and beverages.

The Flow of Food

INTRODUCTION

The job of protecting food continues even after it has been prepared and cooked properly, since microorganisms can still contaminate food before it is eaten. The key to serving safe food is to prevent time-temperature abuse and cross-contamination. Hold, display, and serve food at the right temperature, and handle it safely. People do many things without knowing their actions can lead to contamination. Train employees to serve food properly, and make sure food safety rules are followed.

GENERAL RULES FOR HOLDING FOOD

► **Check the internal temperature of food using a thermometer.** The holding equipment's thermostat measures the temperature of the equipment, not the food.

► **Check the temperature of food at least every four hours.** Food that is not at 135°F (57°C) or higher or at 41°F (5°C) or lower must be discarded. As an alternative, check the temperature every two hours to leave time for corrective action.

► **Establish a policy to ensure that food being held for service will be discarded after a predetermined amount of time.** For example, a policy may state that a pan of veal can be replenished all day as long as it is discarded at the end of the day.

► **Protect food from contaminants with covers or sneeze guards.** Covers help to maintain temperature and keep out contaminants.

► **Prepare food in small batches so it will be used faster.** Do not prepare food any further in advance than necessary to minimize the potential for time-temperature abuse.

Hot Food

► **Potentially hazardous, hot food must be held at an internal temperature of 135°F (57°C) or higher.** (See *Exhibit 9a.*) You can also hold it at an even higher temperature of 140°F (60°C) as an additional safeguard.

Exhibit 9a

Hot Holding

135°F
57°C

Potentially hazardous, hot food must be held at an internal temperature of 135°F (57°C) or higher.

Exhibit 9b

Cold Holding

🚫 ✓ 41°F
5°C

Potentially hazardous, cold food must be held at an internal temperature of 41°F (5°C) or lower.

▶ **Only use hot-holding equipment that can keep food at the proper temperature.**

▶ **Never use hot-holding equipment to reheat food if it is not designed to do so.** Reheat food to 165°F (74°C) for fifteen seconds within two hours, then transfer it to holding equipment. Most hot-holding equipment is incapable of passing food through the temperature danger zone quickly enough.

▶ **Stir food at regular intervals to distribute heat evenly.**

Cold Food

▶ **Potentially hazardous, cold food must be held at an internal temperature of 41°F (5°C) or lower.** (See *Exhibit 9b*.)

▶ **Only use cold-holding equipment that can keep food at the proper temperature.**

▶ **Do not store food directly on ice.** Whole fruit and vegetables and raw, cut vegetables are the only exceptions. Place food in pans or on plates first.

Holding Food Without Temperature Control

Ready-to-eat, potentially hazardous food can be displayed or held for consumption without temperature control for up to four hours under the following conditions:

▶ **Prior to removing the food from temperature control it has been held at 41°F (5°C) or lower, or at 135°F (57°C) or higher.**

▶ **The food contains a label that specifies when the item must be discarded.** The label must reflect a time that is four hours after the item was taken out of temperature control. For example, if potato salad served at a picnic was removed from refrigeration at 12:00 P.M., the time indicated on the label must be 4:00 P.M. because the potato salad must be discarded within four hours.

▶ **The food must be sold, served, or discarded within four hours.**

Before using time as a method of control, check with your regulatory agency for specific requirements in your area.

Apply Your Knowledge

To Serve or Not to Serve

The following food items were held without temperature control. Place an **X** next to the items that were handled properly.

___ **❶** Potato salad is taken out of refrigeration at 10:00 A.M. and is labeled: Discard at 3:00 P.M.

___ **❷** A pan of properly-cooked scrambled eggs is placed in the hot-holding unit at 9:00 A.M., held at 120°F (49°C), and discarded at 11:00 A.M., as the label indicated.

___ **❸** Sliced ham, held at 50°F (10°C) the night before, is held at room temperature for four hours and then discarded as the label indicated.

___ **❹** A caterer delivers several pans of sliced watermelon to be held at room temperature to a picnic and tells employees to discard them after four hours.

For answers, please turn to page 9-17.

SERVING FOOD SAFELY

After handling food safely and cooking it properly, you do not want to risk contamination when serving it.

Kitchen Staff

Train your kitchen staff to follow these procedures for serving food safely.

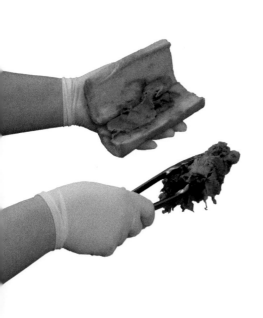

▶ **Store serving utensils properly.** Serving utensils can be stored in the food, with the handle extended above the rim. They can also be placed on a clean, sanitized food-contact surface. Spoons or scoops used to serve food such as ice cream or mashed potatoes can be stored under running water.

▶ **Use serving utensils with long handles.** Long-handled utensils keep the server's hands away from food.

▶ **Use clean and sanitized utensils for serving.** Use separate utensils for each food item, and properly clean and sanitize them after each serving task. Utensils should be cleaned and sanitized at least once every four hours during continuous use.

▶ **Minimize bare-hand contact with food that is cooked or ready-to-eat.** Handle food with tongs, deli sheets, or gloves. Bare-hand contact is allowed in some jurisdictions if the establishment has a verifiable written policy on handwashing procedures. **Check with your regulatory agency for requirements in your area.**

▶ **Practice good personal hygiene.** Wash hands after using the restroom, or after hands have come in contact with anything that may contaminate food.

Servers

Food servers need to be as careful as kitchen staff. If they are not careful, they can contaminate food simply by handling the food-contact surfaces of glassware, dishes, and utensils. Servers should use the following guidelines to serve food safely. (See *Exhibit 9c.*)

▶ **Glassware and dishes should be handled properly.** The food-contact areas of plates, bowls, glasses, or cups should not be touched. Dishes should be held by the bottom or the edge. Cups should be held by their handles, and glassware should be held by the middle, bottom, or stem.

▶ **Glassware and dishes should not be stacked when serving.** The rim or surface of one can be contaminated by the one above it. Stacking china and glassware also can cause it to chip or break.

▶ **Flatware and utensils should be held at the handle.** Store flatware so servers grasp handles, not food-contact surfaces.

▶ **Minimize bare-hand contact with food that is cooked or ready-to-eat.**

▶ **Use ice scoops or tongs to get ice.** Servers should never scoop ice with their bare hands or use a glass since it may chip or break. Ice scoops should always be stored in a sanitary location—not in the ice bin.

▶ **Practice good personal hygiene.** Servers should be neat and clean, and their hair should be pulled back. They should avoid touching their hair or face when serving food, and should refrain from habits such as chewing fingernails or licking their fingers.

▶ **Never use cloths meant for cleaning food spills for any other purpose.** When tables are cleaned between guest seatings, spills should be wiped up with a disposable, dry cloth. The table should then be cleaned with a moist cloth that has been stored in a fresh sanitizer solution.

The Flow of Food

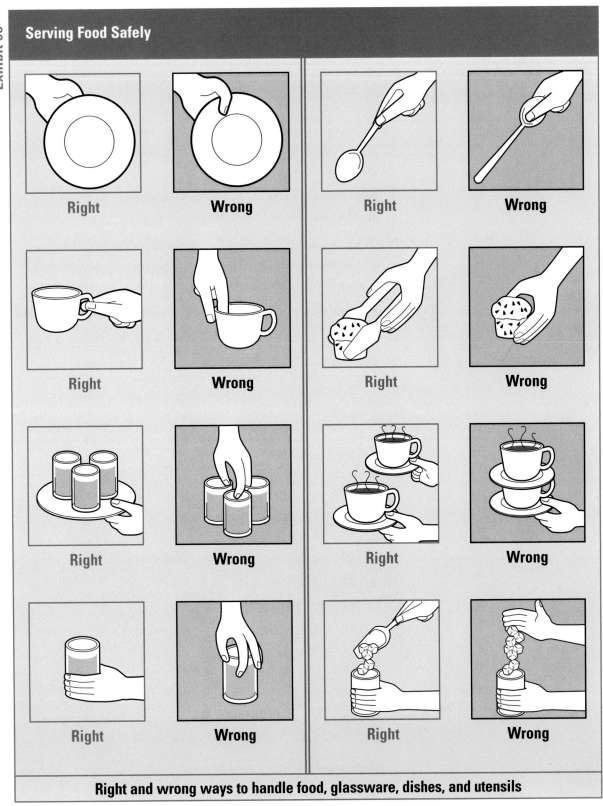

Exhibit 9c

Serving Food Safely

Right and wrong ways to handle food, glassware, dishes, and utensils

Re-serving Food Safely

Servers and kitchen staff should also know the rules about re-serving food that has been previously served to a customer.

▶ **In general, only unopened, prepackaged food, such as condiment packets, wrapped crackers, or wrapped breadsticks, can be re-served.** No food items should be re-served to people at risk, such as nursing-home residents or hospital patients.

▶ **Never re-serve plate garnishes, such as fruit or pickles, to another customer.** Served but unused garnishes must be discarded.

▶ **Never re-serve uncovered condiments.** Do not combine leftovers with fresh food. For example, opened portions of salsa, mayonnaise, mustard, or butter should be thrown away.

▶ **Do not re-serve uneaten bread or rolls to other customers.**

▶ **Linens used to line bread baskets must be changed after each customer.**

Apply Your Knowledge	Serve or Discard?
Place an **R** in the space next to each item if it can be re-served or a **D** if it must be discarded.	___ ❶ Previously served, but untouched basket of bread ___ ❷ Individually wrapped crackers ___ ❸ Unwrapped butter served on a plate ___ ❹ Mustard packets ___ ❺ Bowl of mayonnaise ___ ❻ Unopened, single-serve yogurt cup in a nursing home ___ ❼ Ice used on a food bar ___ ❽ Uncovered cream container left on the table for a day ___ ❾ Fruit garnish **For answers, please turn to page 9-17.**

Exhibit 9d

Self-Service Areas

Assign a trained staff member to monitor food bars and buffets.

Self-Service Areas

Buffets and food bars can be contaminated easily. They should be monitored closely by employees trained in food safety procedures. (See *Exhibit 9d.*) Here are more rules for food bars:

▶ **Protect food on display with sneeze guards or food shields.** These should be fourteen inches above the food counter, and the shield should extend seven inches beyond the food.

▶ **Identify all food items.** Label containers on the food bar. Place the names of salad dressings on ladle handles. (See *Exhibit 9e.*)

▶ **Maintain proper food temperatures.** Keep hot food hot—135°F (57°C) or higher, and cold food cold—41°F (5°C) or lower.

▶ **Replenish food on a timely basis.** Prepare and replenish small amounts at a time. Practice the FIFO method of product rotation.

▶ **Keep raw meat, fish, and poultry separate from cooked and ready-to-eat food.** Use separate displays or food bars for raw and cooked food (for example, at a Mongolian barbecue) to reduce the chance of cross-contamination.

▶ **Do not let customers use soiled plates or silverware for refills.** Assign a staff member to hand out fresh plates for return visits. Post signs with polite tips about food-bar etiquette. Customers can use glassware for refills as long as beverage-dispensing equipment does not come in contact with the rim or interior of the glass.

Exhibit 9e

Food Bars and Buffets

Label containers and ladles.

Apply Your Knowledge	Too Hot to Handle

The Firehouse, a popular buffet, is famous for its Four Alarm Chili. Place an **X** next to each practice that helps ensure the safety of its chili.

___ ❶ The chili is held at an internal temperature of 135°F (57°C).

___ ❷ The temperature of the chili is checked every four hours and if it is not at the proper temperature, the chili is reheated.

___ ❸ Sour cream and other chili condiments are held at 50°F (10°C).

___ ❹ Ladles for serving the chili are stored in the product.

___ ❺ Customers are encouraged to refill their chili bowls.

For answers, please turn to page 9-17.

For answers, please turn to page 9-17.

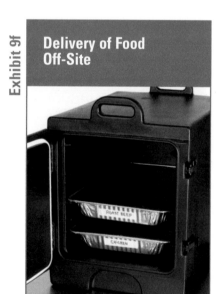

Exhibit 9f

Delivery of Food Off-Site

When transporting food, use rigid, insulated containers capable of maintaining proper temperature.

OFF-SITE SERVICE

Delivery

Many establishments—such as schools, hospitals, caterers, and even restaurants—prepare food at one location and then deliver it to remote sites. The greater the time and distance from the point of preparation to the point of consumption, the greater the risk the food will be exposed to contamination or time-temperature abuse. When transporting food, the following procedures should be followed:

▶ **Use rigid, insulated food containers capable of maintaining food temperatures at 135°F (57°C) or higher or at 41°F (5°C) or lower.** Containers should be sectioned so food does not mix, leak, or spill. They must also allow air circulation to keep temperatures even and should be kept clean and sanitized. (See *Exhibit 9f.*)

▶ **Clean and sanitize the inside of delivery vehicles regularly.**

▶ **Practice good personal hygiene when distributing food.**

▶ **Check internal food temperatures regularly.** Take corrective action if food is not at the proper temperature. If containers or delivery vehicles are not maintaining proper food temperatures at the end of each route, reevaluate the length of delivery routes or the efficiency of the equipment being used.

▶ **Label food with storage, shelf-life, and reheating instructions for employees at off-site locations.**

▶ **Provide food safety guidelines for consumers.**

Catering

Caterers provide food for private parties and events, as well as public and corporate functions. They might bring prepared food, or cook food on-site in a mobile unit, a temporary unit, or in the customer's own kitchen.

▶ **Make sure safe drinking water is available for cooking, warewashing, and handwashing.**

▶ **Ensure that adequate power is available for cooking and holding equipment.**

▶ **Use insulated containers for all potentially hazardous food.** Raw meat should be wrapped and stored on ice. Deliver milk and dairy products in a refrigerated vehicle or on ice.

▶ **Serve cold food in containers on ice or in chilled, gel-filled containers.** If that is not desirable, the food may be held without temperature control according to the guidelines specified in this section.

▶ **Keep raw and ready-to-eat products separate.** For example, store raw chicken separately from ready-to-eat salads.

▶ **Use single-use items.** Make sure customers get a new set of disposable tableware for refills.

▶ **If leftovers are given to customers, provide instructions on how they should be handled.** This may include reheating and storage instructions for the products and shelf-life dates.

▶ **Place garbage-disposal containers away from food-prep and serving areas.**

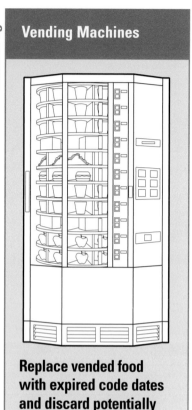

Vending Machines

Replace vended food with expired code dates and discard potentially hazardous food within seven days of preparation.

Exhibit 9g

Vending Machines

Food prepared and packaged for vending machines should be handled with the same care as any other food served to a customer. Vending operators also should protect food from contamination and time-temperature abuse during transport, delivery, and service.

▶ **Keep potentially hazardous food at the right temperature.** It should be held at 41°F (5°C) or lower, or at 135°F (57°C) or higher.

▶ **Machines must have automatic shutoff controls.** These will prevent food from being dispensed if the temperature stays in the danger zone for a specified amount of time.

▶ **Check product shelf life daily.** Replace food with expired code dates. Refrigerated food not used within seven days of preparation must be discarded. (See *Exhibit 9g*.)

▶ **Dispense potentially hazardous food in its original container.** Fresh fruit with edible peels should be washed and wrapped before being put in a machine.

The Flow of Food

Apply Your Knowledge

A Case in Point

❶ What did Jill do wrong?

❷ What should have been done?

For answers, please turn to page 9-17.

Jill, a line cook on the morning shift at Memorial Hospital, was busy helping the kitchen staff put food on display for lunch in the hospital cafeteria. Ann, the kitchen manager who usually supervised lunch in the cafeteria, was at an all-day seminar on food safety. Jill was responsible for making sure meals were trayed and put into food carts for transport to the patients' rooms. The staff also packed two dozen meals each day for a neighborhood group that delivered them to homebound elderly people.

First, Jill looked for insulated food containers for the delivery meals. When she could not find them, she loaded the meals into cardboard boxes she found near the back door, knowing the driver would be there soon to pick them up. To help the cafeteria staff, Jill filled a baine with soup by dipping a two-quart measuring cup into the stockpot and pouring it into the baine. She carried the baine out to the cafeteria, put it into the steam table, and turned it on low.

The lunch hour was hectic. The cafeteria was busy, and the staff had many patient meals to tray and deliver. Halfway through lunch, a cashier came back to the kitchen to tell Jill that the salad bar needed replenishing. Since she was busy, Jill asked a kitchen employee to take pans of prepared ingredients out of the refrigerator and put them on the salad bar. When she looked up a few moments later, she saw the kitchen employee send away two children who were eating carrot sticks from the salad bar.

With lunch almost over, Jill breathed a sigh of relief. She moved down the cafeteria serving line, checking food temperatures. One of the casseroles was about 130°F (54°C). Jill checked the water level in the steam table and turned up the thermostat, then went to clean up the kitchen and finish her shift.

SUMMARY

Safe foodhandling does not stop once food is properly prepared and cooked. To make sure the food you serve is safe, you must continue to protect it from time-temperature abuse and contamination until it is eaten. When holding potentially hazardous food for service, keep hot food hot—135°F (57°C) or higher, and cold food cold—41°F (5°C) or lower. Check the internal temperature of food being held at least every four hours, and discard it if it is not at the proper temperature. Protect food from contaminants by using covers and lids, and establish policies for discarding food held for service after a predetermined amount of time. If food will be held without temperature control, label it with a discard time, and sell, serve, or discard it after four hours. Use clean, sanitized utensils to serve food and minimize bare-hand contact with cooked and ready-to-eat food.

Make sure all employees practice good personal hygiene. Train them to avoid cross-contamination when handling service items and tableware. Teach them about the potential hazards posed by re-serving plate garnishes, breads, or open dishes of condiments.

Customers can unknowingly contaminate food in self-service areas. Post signs to communicate self-service rules, and station employees in these areas to ensure compliance. Protect food in food bars and buffets with sneeze guards, and make sure equipment can hold food at the proper temperature. Take special precautions when preparing, delivering, or serving food off-site. Learn and follow all the regulations in your jurisdiction.

Apply Your Knowledge   Multiple-Choice Study Questions

Use these questions to test your knowledge of the concepts presented in this section.

1. Which of the following is an unsafe serving practice?
   A. Stacking plates of food before serving them to the customer
   B. Holding flatware by the handles when setting a table
   C. Serving soup with a long-handled ladle
   D. Holding glassware by the stem

2. Which of the following statements about serving utensils is *not* true?
   A. They should be cleaned and sanitized at least once every four hours during continuous use.
   B. They can be used to handle more than one food item at a time.
   C. They can be stored in the food with the handle extended above the rim of the container.
   D. They must be cleaned and sanitized after each task.

3. To keep vended food safe, you should
   A. leave fresh fruit with edible peels unwrapped.
   B. discard food within fourteen days of preparation.
   C. dispense potentially hazardous food in its original container.
   D. ensure that cold, potentially hazardous food is kept at 50°F (10°C) or lower.

4. To hold cold food safely, you should
   A. store it directly on ice.
   B. store it at 41°F (5°C) or lower.
   C. stir it regularly.
   D. leave it uncovered.

Continued on next page...

Apply Your Knowledge   **Multiple-Choice Study Questions** *continued*

5. Which of the following is an acceptable serving practice at a self-service bar?
   A. Holding hot, potentially hazardous food at 120°F (49°C)
   B. Storing raw meat next to ready-to-eat food
   C. Allowing customers to use the same plate for a return trip to the self-service bar
   D. Allowing customers to re-use glassware for beverage refills

6. Hot, potentially hazardous food should be held at an internal temperature of
   A. 135°F (57°C) or higher.
   B. 130°F (54°C) or higher.
   C. 120°F (49°C) or higher.
   D. 110°F (43°C) or higher.

7. All of the following are conditions for allowing food to be held without temperature control *except*
   A. the food must contain a label specifying when it must be discarded.
   B. gloves must be worn when handling the food.
   C. the local jurisdiction must allow food to be held without temperature control.
   D. the food must be sold, served, or discarded within four hours.

For answers, please turn to page 9-17.

Apply Your Knowledge   Answers

Page	Activity

9-2   Test Your Food Safety Knowledge

1. True   2. False   3. False   4. True   5. True

9-5   To Serve or Not to Serve

2. Only the scrambled eggs were handled properly.

9-8   Serve or Discard?

1. D	3. D	5. D	7. D	9. D
2. R	4. R	6. D	8. D	

9-10   Too Hot to Handle

1,4

9-13   A Case in Point

❶ Here is what Jill did wrong: Ⓐ She packed the deliveries in cardboard boxes instead of rigid, insulated carriers. Ⓑ She used the wrong utensil to fill the soup baine. Ⓒ She failed to make sure that the internal temperature of the food on the steam table was checked at least every four hours. This would have alerted her to the fact that the steam table was not maintaining the proper temperature and that the casserole was in the temperature danger zone. ❷ Jill should have done the following: Ⓐ She should have kept the delivery meals in a hot-holding cabinet or left the food in a steam table until suitable containers were found or the driver arrived. Ⓑ She should have used a long-handled ladle, which would have kept her hands away from the soup, preventing possible contamination, as she ladled it out. Ⓒ She should have made sure that an employee was assigned to monitor the food bar to ensure that customers, such as the children, followed proper etiquette. Ⓓ She should have discarded the casserole and any other food that was not at the right temperature, since she did not know how long the food was in the TDZ.

9-15   Multiple-Choice Study Questions

1. A	3. C	5. D	7. B
2. B	4. B	6. A	

Food Safety Systems

Inside this section:
▶ Food Safety Programs
▶ Active Managerial Control
▶ HACCP
▶ When a HACCP Plan is Required
▶ Crisis Management

After completing this section, you should be able to:
▶ Identify how active managerial control can impact food safety.
▶ Identify HACCP principles for preventing foodborne illness.
▶ Implement HACCP principles when applicable.
▶ Identify when a HACCP plan is required.
▶ Implement a crisis management program.
▶ Cooperate with regulatory agencies in the event of a foodborne-illness investigation.

Apply Your Knowledge	Test Your Food Safety Knowledge

Check to see how much you know about the concepts in this section. Use the page references provided to explore the topic in each question.

❶ **True or False:** Active managerial control focuses on taking action to control three foodborne-illness risk factors identified by the CDC. *(See page 10-4.)*

❷ **True or False:** Purchasing fish from local fishermen would be considered a risk under active managerial control. *(See page 10-3.)*

❸ **True or False:** Cooking chicken to a minimum internal temperature of 165°F (74°C) for fifteen seconds would be an appropriate critical limit. *(See page 10-8.)*

❹ **True or False:** A critical control point (CCP) is a point in the flow of food where a hazard can be prevented, eliminated, or reduced to safe levels. *(See page 10-8.)*

❺ **True or False:** An establishment that cures food must have a HACCP plan. *(See page 10-11.)*

For answers, please turn to page 10-20.

CONCEPTS

▶ **Food safety management system:** A group of programs and procedures designed to prevent foodborne illness by actively controlling hazards throughout the flow of food.

▶ **Active managerial control:** Proactive approach for addressing the five most common risk factors responsible for foodborne illness as identified by the CDC. By continuously monitoring and verifying procedures responsible for preventing these risks establishments can ensure they are being controlled.

▶ **Hazard Analysis Critical Control Point system (HACCP):** System based on the idea that if significant biological, chemical, or physical hazards are identified at specific points within a product's flow through the operation, they can be prevented, eliminated, or reduced to safe levels.

INTRODUCTION

In Sections 6 through 9, you learned how to handle food safely as it flows through the establishment from purchasing and receiving, through storing, preparing, cooking, holding, cooling, reheating, and serving. This accumulated knowledge will help you develop a food safety management system to prevent foodborne illness by actively controlling hazards throughout the flow of food. In this chapter, you will be introduced to the fundamentals for developing this type of system.

FOOD SAFETY PROGRAMS

A food safety management system must be built on a solid foundation of programs that supports your efforts to minimize the risk of foodborne illness. These programs include (see *Exhibit 10a* on the next page):

▶ Personal hygiene (including restriction and exclusion policies for sick employees)

▶ Facility design

▶ Supplier selection and specification

▶ Cleaning and sanitation

▶ Equipment maintenance

▶ Manager and employee food safety training

The development and maintenance of these programs are crucial for addressing the five most common risk factors responsible for foodborne illness as identified by the CDC. The factors are:

▶ Purchasing food from unsafe sources

▶ Failing to cook food adequately

▶ Holding food at improper temperatures

▶ Using contaminated equipment

▶ Poor personal hygiene

Food Safety Programs

The following programs must be in place for a food safety management system to be effective.

Proper personal hygiene program

Proper facility-design program

Supplier selection and specification programs

Proper cleaning and sanitation programs

Appropriate equipment-maintenance programs

Food safety training programs

ACTIVE MANAGERIAL CONTROL

Active managerial control is a proactive, rather than reactive, approach to addressing the CDC's risks. By continuously monitoring and verifying procedures responsible for preventing these risks, you will ensure they are being controlled.

Approach

There are specific steps that should be taken when developing a food safety management system using active managerial control.

Step 1: Establish the necessary food safety programs, and support them through your standard operating procedures.

Active Managerial Control

Managers must monitor policies and procedures to ensure that they are being followed.

Step 2: Consider the five risk factors as they apply throughout the flow of food, and identify the potential breakdowns that could impact food safety.

Step 3: If necessary, revise policies and procedures to prevent these breakdowns from occurring.

Step 4: Monitor the policies and procedures to ensure they are being followed. (See *Exhibit 10b*.)

Step 5: Verify that the policies and procedures you have established are actually controlling the risk factors. Use feedback from internal sources (records, temperature logs, and self-inspections) and external sources (health inspection reports, customer comments, and quality assurance audits) to adjust your policies and procedures in order to continuously improve the system.

For an example of active managerial control in action, see *Exhibit 10c*.

Example of Active Managerial Control

A full-service seafood restaurant chain has identified buying seafood from an unsafe source as a risk in their establishments during purchasing. To avoid buying unsafe product, management has decided to do the following:

- ▶ Develop criteria for creating an approved list of inspected vendors.

- ▶ Create a policy stating that seafood may only be purchased from vendors on this list.

- ▶ Periodically monitor invoices and deliveries to ensure they are coming from approved vendors.

- ▶ Regularly verify that the criteria set for the vendors are still appropriate for controlling the risk.

- ▶ Review the policies and procedures when a breakdown occurs, to determine why, and change them to ensure it does not happen again.

Apply Your Knowledge

Five of these unsafe practices are responsible for the majority of foodborne illnesses. Place an **X** next to each one.

Which is Worse?

___ ❶ Cooling food improperly

___ ❷ Failing to follow FIFO when storing food

___ ❸ Purchasing food from unsafe sources

___ ❹ Failing to cook food adequately

___ ❺ Date-marking food improperly

___ ❻ Failing to implement an integrated pest management program

___ ❼ Using contaminated equipment

___ ❽ Poor personal hygiene

___ ❾ Thawing food at improper temperatures

___ ❿ Holding food at improper temperatures

For answers, please turn to page 10-20.

HACCP

A food safety management system may also include a Hazard Analysis Critical Control Point (HACCP) system. A HACCP (pronounced *Hass-ip*) system is based on the idea that if significant biological, chemical, or physical hazards are identified at specific points within a product's flow through an operation, they can be prevented, eliminated, or reduced to safe levels.

To be effective, a HACCP system must be based on a written plan that is specific to each facility's menu, customers, equipment, processes, and operations. A HACCP plan is based on the seven basic principles outlined by the National Advisory Committee on Microbiological Criteria for Foods.

Approach

The HACCP principles are seven sequential steps that outline how to create a HACCP plan. Since each principle builds on the information gained from the previous principle, when developing your plan you must consider all seven principles in order.

In general terms:

▶ Principles One and Two help you identify and evaluate your hazards.

▶ Principles Three, Four, and Five help you establish how you will control those hazards.

▶ Principles Six and Seven help you maintain both your HACCP plan and system and verify their effectiveness.

The Seven HACCP Principles

Principle One: Conduct a Hazard Analysis

Identify and assess potential hazards in the food you serve by taking a look at how it is processed, or flows through your establishment. Many types of food are processed similarly. The most common processes include:

▶ Preparing and serving without cooking

▶ Preparing and cooking for same-day service

▶ Preparing, cooking, holding, cooling, reheating, and serving—which is also called complex food preparation

Once common processes have been identified, you can determine where food safety hazards are likely to occur for each one. Hazards include:

▶ Bacterial, viral, or parasitic contamination

▶ Contamination by cleaning compounds, sanitizers, and allergens

▶ General physical contamination

Principle Two: Determine Critical Control Points (CCPs)

Find the points in the process where the identified hazard(s) can be prevented, eliminated, or reduced to safe levels. These are the critical control points (CCPs). Depending on the process, there may be more than one CCP.

Principle Three: Establish Critical Limits

For each CCP, establish minimum and maximum limits that must be met to prevent or eliminate the hazard, or to reduce it to a safe level.

Principle Four: Establish Monitoring Procedures

Once critical limits have been established, determine the best way for your operation to check them to make sure they are consistently met. Identify who will monitor them and how often.

Principle Five: Identify Corrective Actions

Identify steps that must be taken when a critical limit is not met. These steps should be determined in advance.

Principle Six: Verify that the System Works

Determine if the plan is working as intended. Plan to evaluate on a regular basis your monitoring charts, records, how you performed your hazard analysis, etc., and determine if your plan adequately prevents, reduces, or eliminates identified hazards.

Principle Seven: Establish Procedures for Record Keeping and Documentation

Maintain your HACCP plan. Keep records obtained while performing monitoring activities, whenever a corrective action is taken, when equipment is validated (checked to make sure it is in good working condition), and when working with suppliers (i.e., shelf-life studies, specifications, challenge studies, etc.). Also keep all documentation created while you were developing the plan.

To follow the development of a HACCP plan, see *Exhibit 10d* for an example involving Enrico's, a full-service Italian restaurant.

Exhibit 10d

Development of a HACCP plan at Enrico's

**Principle One—
Conduct a Hazard
Analysis**

Enrico's management team began by conducting a hazard analysis. Looking at their menu, they noted that several of their dishes—including *Chicken Breast alla Parmigiana* and *Pepper Steak*—used the same process of receiving, storage, preparation, cooking, and same-day service.

The team determined that several biological hazards were most likely to affect the food prepared by this process. In *Chicken Breast alla Parmigiana*, *Salmonella* spp. and *Campylobacter* spp. are the most likely biological hazards, while shiga toxin-producing *E. coli* could affect the *Pepper Steak*.

**Principle Two—
Determine Critical
Control Points**

Enrico's management team identified cooking as a critical control point for this process. While proper food safety practices must be followed throughout the food's flow, proper cooking is the only step that will eliminate or reduce the identified hazards to safe levels. Since the food was prepared for same-day service, it was the only CCP identified.

**Principle Three—
Establish Critical
Limits**

Since cooking was identified as a CCP for the process, the team determined that their critical limit for *Chicken Breast alla Parmigiana* would be cooking the chicken to a minimum internal temperature of 165°F (74°C) for fifteen seconds. For *Pepper Steak,* the beef must be cooked to a minimum internal temperature of 145°F (63°C) for fifteen seconds. They determined that the critical limits would be met by placing the chicken in a convection oven set to 350°F (177°C) and cooking it for forty-five minutes, and by sautéing the beef to the required temperature.

**Principle Four—
Establish Monitoring
Procedures**

Since *Chicken Breast alla Parmigiana* is cooked to order, Enrico's management team chose to monitor their critical limit by inserting a clean and sanitized thermocouple probe into the thickest part of each breast. Employees were instructed to record the readings in a temperature log. The team chose to monitor the critical limit of the *Pepper Steak* by taking sample temperatures of the beef.

Continued on next page...

Development of a HACCP plan at Enrico's *continued*

**Principle Five—
Identify Corrective
Actions**

In the event that the chicken breast or the beef has not reached its respective critical limit, Enrico's employees have been instructed to keep cooking the food until it does. This is the corrective action, which is recorded in the temperature log.

**Principle Six—
Verify That the
System Works**

Enrico's management team checked their temperature logs on a weekly basis to verify that their critical limits were being met. They noticed that occasionally the chicken breast was not meeting its critical limit, but that the appropriate corrective action was being taken to ensure that the chicken was properly cooked.

The HACCP plan was reevaluated six months after implementation. The reevaluation revealed that the beef consistently met the critical limit set by the management team; however, the chicken routinely failed to meet its set critical limit. Upon reevaluating the cooking process, it was discovered that Enrico's vendor had started supplying a slightly larger chicken breast. This caused the chicken to be undercooked, given the equipment and established cooking parameters. Enrico's cooking process was adjusted to account for the larger chicken breast.

**Principle Seven—
Establish Procedures
for Record Keeping
and Documentation**

Enrico's management team determined that time-temperature logs should be kept for three months and that receiving invoices should be kept for sixty days. They used this documentation to support and revise their HACCP plan as needed, such as reflecting the change in the chicken's cooking process.

Apply Your Knowledge	It's the Principle of the Thing

Identify the HACCP term described in each statement, and place the letter of the term in the space provided.

___ ❶ Checking to see if critical limits are being met

___ ❷ Retention of documents obtained when creating and implementing the HACCP plan

___ ❸ Assessing risk within the flow of food

___ ❹ Specific places in the flow of food where a hazard can be prevented, eliminated, or reduced to safe levels

___ ❺ Predetermined step taken when a critical limit is not met

___ ❻ Minimum and maximum boundaries that must be met to prevent a hazard

___ ❼ Determining if the HACCP plan is working as intended

A. Hazard analysis
B. Critical control points
C. Critical limits
D. Monitoring
E. Corrective action
F. Verification
G. Record keeping and documentation

For answers, please turn to page 10-20.

WHEN A HACCP PLAN IS REQUIRED

The National Restaurant Association and the FDA recommend that all restaurants and foodservice establishments, no matter how large or small, develop and implement a food safety management system. Establishments that perform the following activities, however, must have a HACCP plan in place:

▶ Smoke or cure food as a method of food preservation

▶ Use food additives as a method of food preservation

▶ Package food using a reduced-oxygen packaging method

▶ Offer live, molluscan shellfish from a display tank

▶ Custom-process animals for personal use

▶ Package unpasteurized juice for sale to the consumer without a warning label

CRISIS MANAGEMENT

A food safety system is designed to help you take steps to ensure that the food you serve is safe. Despite your best efforts, however, a foodborne-illness outbreak can occur in your establishment at any time. How you respond when that happens can determine whether or not you end up in the middle of a crisis.

The basis of a successful crisis management program is a written plan that identifies the resources required and procedures that must be followed to handle crises. The time to prepare for a crisis is before one occurs. There is no off-the-shelf disaster plan that works for every establishment. Each plan must meet its operation's individual needs.

Developing a Plan

When developing your plan, start by stating the basic objectives of the plan. These usually include meeting the immediate needs of the operation and keeping the business viable. Then, include the level of detail for the plan. This may consist of a checklist with brief step-by-step procedures to follow or a full-scale plan covering specific tasks, roles, and resources. You should also prepare specific procedures for developing, updating, and distributing the plan.

There are a number of steps you can take to prepare for the possibility of a crisis.

▶ **Develop a crisis management team.** In large, multi-unit operations, the team may include the president and senior managers from finance, operations, marketing, franchising, human resources, public relations, and training departments. Actual crisis-management teams are usually much smaller and may vary in makeup depending on the establishment and the situation. An independent establishment's team might include the owner, general manager, and chef. (See *Exhibit 10e.*)

Develop a Crisis Management Team

Develop a team so that every task and role in your plan will be supported in the event of a crisis.

Exhibit 10e

▶ **Identify potential crises.** While the greatest threat to customers may be from foodborne illness, do not forget that other crises can include food security threats, robberies, severe weather, fire, or some other trauma.

▶ **Develop simple instructions on what to do in each type of crisis.** In a foodborne-illness outbreak, steps include isolating the suspect food, obtaining samples of the suspect food, preventing further sale of the food, excluding suspect employees from handling food, and contacting the local health department.

▶ **Assemble a contact list of names and numbers, and post it by the phones.** The list should include all crisis-management team members and outside resources, such as police, fire and health departments, testing labs, issues experts, and management or headquarters personnel.

▶ **Develop a crisis communication plan.** It should include:

  ▶ A list of media responses or a Q&A sheet suggesting what to say in the event of each type of crisis identified. Create sample press releases that can be tailored quickly to each incident.

  ▶ A list of media contacts to call for press conferences or news briefings. Include a media-relations plan with do's and don'ts for dealing with the media.

  ▶ How to communicate with employees. Possibilities include shift meetings, email, a telephone tree, etc.

▶ **Assign and train a spokesperson to handle media relations.** Appoint a single spokesperson to handle all media queries and communications. Designating a point person usually results in more consistent messages and allows you to control media access to your staff. The spokesperson should be familiar with interview skills so that he or she knows what to expect and how to respond. Crisis situations can be very stressful, and training will enable your spokesperson to handle it better. Make sure all of your staff knows who the spokesperson is, and instruct them to direct questions to that person.

▶ **Assemble a crisis kit for the establishment.** The kit can be in the form of a three-ring notebook or binder enclosing the plan's materials. Keep the kit in an accessible place, such as the manager's or chef's office.

▶ **Test the plan by running a simulation.** Hire a public relations or consulting firm with crisis-management experience to enact a crisis and test your team's readiness. In most cases, the firm will design a simulated crisis that will be as close as possible to what could happen in a real-life situation.

Crisis Response

You may be able to avert a crisis by responding quickly when you do receive customer complaints. Take all customer complaints seriously. Express your concern and be sincere, but do not admit responsibility or accept liability. Listen carefully and promise to investigate and respond.

With legal guidance, consider developing an incident report to help you through the process. Questions to ask include:

▶ What did you eat and drink at our establishment and when?

▶ When did you become ill? What were the symptoms, and how long did you experience them?

▶ Did you eat anything else before or after eating at our establishment? What and where? Who else ate the same food, and did they become ill?

▶ Did you seek medical attention? Where and how soon after becoming ill? What diagnosis and treatment did you receive?

Evaluate the complaint. If more than one person has complained, you have the potential for a crisis on your hands. Take steps to control the situation and reassure customers that you are doing everything you can to identify and fix the problem. At this point, call your crisis team together and implement your plan.

▶ Direct the team to gather information, plan courses of action, and manage events as they unfold.

▶ Work with, not against, the media. Be as proactive as you can, as early as you can. Make sure the spokesperson is fully informed before arranging a press conference. Contacting the media before they contact you helps you to control what the media reports. Stick to the facts, and be as honest as possible. If you do not have all the facts, say so, and let the media know that you will communicate them as soon as you do know. Keep a cool head and do not be defensive. The easiest way to magnify or prolong a crisis is to deny, lie, or change your story.

▶ Show concern and be sincere. If health officials have confirmed that your establishment is the source of the illness, accept responsibility. Accepting responsibility is not the same as admitting liability. While customers may have become ill from eating food in your operation, the cause may have been beyond your control and not your fault. If you do not express your concern, and mean it, you will lose credibility with the public, not just customers.

▶ Communicate the information directly to all of your key audiences. Do not depend on the media to relay all the facts. Tell your side of the story to employees, customers, stockholders, and the community. Use newsletters, a Web site, flyers, and newspaper or radio advertising.

▶ Fix the problem and communicate what you have done both to the media and to your customers. Each time you take a step to resolve the problem, let the media know. Hold briefings when you have news, and go into each briefing or press conference with an agenda. Take control. Do not simply respond to questions.

Post Crisis Assessment

Once a crisis is over, it is important to determine the causes and effects of the crisis so that your establishment can implement changes to take advantage of the lessons learned. Some things to evaluate include:

▶ Obstacles faced in returning to normal operations

▶ How communication with employees and customers was handled

▶ Assessment of the damages from the crisis

▶ Overall assessment of the crisis and your response

A foodborne-illness outbreak has the potential to damage your business beyond repair. Investing time and resources in a crisis management plan can ensure against that. Remember these three key rules of crisis management:

❶ Take steps to prevent a crisis from occurring by practicing good food safety habits.

❷ Prepare for the possibility of a crisis by developing contingency plans.

❸ If a crisis does occur, take control of the situation. Use your plan to manage the crisis thoughtfully, honestly, and as quickly as possible.

SUMMARY

A food safety management system will help you prevent foodborne illness by controlling hazards throughout the flow of food.

Active managerial control focuses on establishing policies and procedures to control five common risk factors responsible for foodborne illness: purchasing food from unsafe sources, failing to cook food adequately, holding food at improper temperatures, using contaminated equipment, and poor personal hygiene. The polices and procedures that an establishment puts in place, or revision of existing ones, will be the result of a careful analysis of potential breakdowns related to these five risk factors as they apply throughout the flow of food. Once procedures have been implemented, they must be continuously monitored, and the

system must be verified to ensure that the procedures put in place are controlling the identified risks.

A HACCP (Hazard Analysis Critical Control Point) system focuses on identifying specific points within a product's flow through the operation that are essential to prevent, eliminate, or reduce biological, chemical, or physical hazards to safe levels. To be effective, a HACCP system must be based on a plan specific to a facility's menu, customers, equipment, processes, and operation. The HACCP plan is developed following seven sequential principles—essential steps for building a food safety system.

First, the establishment must identify and assess potential hazards in the food they serve by taking a look at how it is processed. Once common processes have been identified, they can determine where food safety hazards are likely to occur for each one. The establishment must then identify points where they can be prevented, eliminated, or reduced to safe levels. These are the critical control points (CCPs). Next, the establishment must determine minimum and maximum limits that must be met for each CCP to prevent, eliminate, or reduce the hazard. The establishment must determine how they will monitor the CCPs they have identified and what actions will be taken when critical limits have not been met. Finally, the establishment must find ways to verify that the HACCP system is working, and establish procedures for record keeping and documentation.

A food safety system is designed to help you take steps to ensure that the food you serve is safe. Despite your best efforts, however, a foodborne-illness outbreak can occur in your establishment. The time to prepare for a crisis is before one occurs. The key is to start with a written plan that identifies the resources required and procedures that must be followed to handle the crisis. To prepare for a crisis: develop a crisis management team, spell out specific instructions for handling the crisis, develop a crisis communication plan, and assign and train a media spokesperson. When receiving customer complaints, listen carefully, express concern, and be sincere. Do not admit responsibility, but promise to investigate and respond. With legal advice, consider developing an incident report to guide you through the process. Call your crisis management team together and implement your plan.

Apply Your Knowledge

Use these questions to test your knowledge of the concepts presented in this section.

Multiple-Choice Study Questions

1. The temperature of a roast is checked to see if it has met its critical limit of 145°F (63°C). This is an example of which HACCP principle?
   A. Verification
   B. Monitoring
   C. Record keeping
   D. Hazard analysis

2. The temperature of a pot of beef stew is checked during holding. The stew has not met the critical limit of 135°F (57°C) and is discarded according to house policy. This is an example of which HACCP principle?
   A. Monitoring
   B. Corrective action
   C. Hazard analysis
   D. Verification

3. Al's Big Burgers makes some of the juiciest burgers in town. Every hamburger is cooked from fresh ground beef to a minimum internal temperature of 150°F (66°C) for fifteen seconds and then dressed with all the trimmings. Al's establishment is at risk of
   A. using contaminated equipment.
   B. failing to receive food properly.
   C. failing to cook food adequately.
   D. failing to store food properly.

4. Which of the following risks is *not* commonly responsible for foodborne illness?
   A. Failing to cook food adequately
   B. Failing to thaw food properly
   C. Failing to purchase food from safe sources
   D. Failing to hold food at the proper temperature

5. Which of the following is *not* a corrective action?
   A. Continuing to cook a hamburger until it reaches a minimum internal temperature of 155°F (68°C) for fifteen seconds
   B. Discarding cooked chicken that has been held at 120°F (49°C) for five hours
   C. Sanitizing a prep counter before starting a new task
   D. Rejecting a shipment of oysters received at 55°F (13°C) that will be served raw

Continued on next page...

6. Which of the following programs should be in place before you begin developing your food safety system?
   A. Personal hygiene program
   B. Incentive program
   C. Workplace accident prevention program
   D. None of the above

7. The purpose of a food safety management system is to
   A. identify the proper methods for receiving food.
   B. identify and control possible hazards throughout the flow of food.
   C. keep the establishment pest free.
   D. identify faulty equipment within the establishment.

8. A chef sanitized his thermometer probe and checked the temperature of a baine of minestrone soup being held in a hot-holding unit. The temperature was 120°F (49°C), which did not meet the establishment's critical limit of 135°F (57°C). He recorded the temperature in the log and reheated the soup to 165°F (74°C) for fifteen seconds. Which was the corrective action?
   A. Sanitizing the thermometer probe
   B. Taking the temperature of the soup
   C. Reheating the soup
   D. Recording the temperature of the soup in the temperature log

9. A HACCP plan is required when an establishment
   A. serves raw shellfish.
   B. serves undercooked ground beef.
   C. uses mushrooms that have been picked in the wild.
   D. packages unpasteurized juice for sale to consumers without a warning label.

For answers, please turn to page 10-20.

Apply Your Knowledge   Answers

Page	Activity

10-2   Test Your Food Safety Knowledge

1. False   2. True   3. True   4. True   5. True

10-6   Which is Worse?

3, 4, 7, 8, and 10 should be marked.

10-11   It's the Principle of the Thing

1. D   3. A   5. E   7. F

2. G   4. B   6. C

10-18   Multiple-Choice Study Questions

1. B   4. B   7. B

2. B   5. C   8. C

3. C   6. A   9. D

Apply Your Knowledge   Notes

Unit 3

Sanitary Facilities and Pest Management

Sanitary Facilities and Pest Management

Inside this section:

▶ **Sanitary Facilities and Equipment**
  ▷ Materials for Interior Construction
  ▷ Considerations for Specific Areas of the Facility
  ▷ Sanitation Standards for Equipment

  ▷ Installing and Maintaining Kitchen Equipment
  ▷ Utilities

▶ **Cleaning and Sanitizing**
  ▷ Cleaning Agents
  ▷ Sanitizing
  ▷ Machine Warewashing
  ▷ Cleaning and Sanitizing in a Three-Compartment Sink
  ▷ Cleaning and Sanitizing Equipment
  ▷ Cleaning Nonfood-Contact Surfaces

  ▷ Tools For Cleaning
  ▷ Storing Utensils, Tableware, and Equipment
  ▷ Using Hazardous Materials
  ▷ Implementing a Cleaning Program

▶ **Integrated Pest Management**
  ▷ The Integrated Pest Management Program
  ▷ Denying Pests Access to the Establishment
  ▷ Denying Pests Food and Shelter

  ▷ Identifying Pests
  ▷ Working with a Pest Control Operator
  ▷ Using and Storing Pesticides

After completing this section, you should be able to:

▶ **Sanitary Facilities and Equipment**
  ▷ Identify when a plan review is required.
  ▷ Identify organizations that certify equipment that meets sanitation standards.
  ▷ Identify characteristics of an appropriate food-contact and nonfood-contact surface.
  ▷ Identify the requirements for installing stationary and mobile equipment.
  ▷ Recognize the importance of maintaining equipment.
  ▷ Identify and prevent cross-connection and backflow.
  ▷ Identify requirements for handwashing facilities including appropriate locations and numbers.

  ▷ Identify the proper response to a waste-water overflow.
  ▷ Recognize the importance of properly installing and maintaining grease traps.
  ▷ Identify potable water sources and testing requirements.
  ▷ Identify lighting-intensity requirements for different areas of the establishment.
  ▷ Identify methods for preventing lighting sources from contaminating food.
  ▷ Identify methods for preventing ventilation systems from contaminating food and food-contact surfaces.

Continued on the next page...

▶ Sanitary Facilities and Equipment *continued*

▷ Identify requirements for storing indoor and outdoor waste.

▷ Identify proper methods for cleaning waste receptacles.

▷ Recognize the need for frequent waste removal to prevent odor and pest problems.

▷ Identify characteristics of appropriate flooring for food establishments.

▷ Recognize the importance of complying with ADA requirements for facility design.

▷ Recognize the importance of keeping physical facilities in proper repair.

▷ Identify requirements for warewashing facilities.

▶ Cleaning and Sanitizing

▷ Explain the difference between cleaning and sanitizing.

▷ Identify approved sanitizers.

▷ Identify factors affecting the efficiency of sanitizers (i.e., time, temperature, concentration, water hardness, and pH).

▷ Use the appropriate test kit for each sanitizer.

▷ Follow the requirements for frequency of cleaning and sanitizing food-contact surfaces.

▷ Properly clean and sanitize items in a three-compartment sink.

▷ Properly clean and sanitize food-contact surfaces.

▷ Properly clean nonfood-contact surfaces.

▷ Identify proper machine-warewashing techniques.

▷ Identify storage requirements for poisonous or toxic materials.

▷ Dispose of poisonous or toxic materials according to legal requirements.

▷ Follow the legal requirements for the use of poisonous or toxic material in a food establishment.

▷ Properly store tools, equipment, and utensils that have been sanitized.

▶ Integrated Pest Management

▷ Identify requirements of an integrated pest management program.

▷ Differentiate between pest prevention and pest control.

▷ Identify ways to prevent pests from entering the facility.

▷ Identify the signs of pest infestation and/or activity.

▷ Identify requirements for applying pesticides.

▷ Identify proper storage requirements for pesticides and pest-application products.

Apply Your Knowledge

Check to see how much you know about the concepts in this section. Use the page references provided to explore the topic in each question.

Test Your Food Safety Knowledge

① **True or False:** There must be a minimum of twenty foot-candles of light (220 lux) in a food-preparation area. *(See page 11-16.)*

② **True or False:** When mounted on legs, stationary equipment must be at least two inches off the floor. *(See page 11-11.)*

③ **True or False:** Cleaning reduces the number of microorganisms on a surface to safe levels. *(See page 11-18.)*

④ **True or False:** Utensils cleaned and sanitized in a three-compartment sink should be dried with a clean towel. *(See page 11-23.)*

⑤ **True or False:** A strong oily odor may indicate the presence of cockroaches. *(See page 11-34.)*

For answers, please turn to page 11-42.

CONCEPTS

▶ **Air gap:** Air space used to separate a water-supply outlet from any potentially contaminated source. The air space between the floor drain and the drain pipe of a sink is an example. An air gap is the only completely reliable method for preventing backflow.

▶ **Backflow:** Unwanted reverse flow of contaminants through a cross-connection into a potable water system. It occurs when the pressure in the potable water supply drops below the pressure of the contaminated supply.

▶ **Cross-connection:** Physical link through which contaminants from drains, sewers, or other waste-water sources can enter a potable water supply. A hose connected to a faucet and submerged in a mop bucket is an example.

▶ **Potable water:** Water that is safe to drink.

▶ **Vacuum breaker:** Device for preventing the backflow of contaminants into a potable water system.

▶ **Cleaning:** Process of removing food and other types of soil from a surface such as a countertop or plate.

▶ **Sanitizing:** Process of reducing the number of microorganisms on a clean surface to safe levels.

▶ **Heat sanitizing:** Using heat to reduce the number of microorganisms on a clean surface to safe levels. The most common way to heat-sanitize tableware, utensils, or equipment is to submerge them in or spray them with hot water.

▶ **Chemical sanitizing:** Using a chemical solution to reduce the number of microorganisms on a clean surface to safe levels. Items can be sanitized by immersing them in a specific concentration of sanitizing solution for a required period of time, or by rinsing, swabbing, or spraying the items with a specific concentration of sanitizing solution.

▶ **Sanitizer:** Chemical used to sanitize. Chlorine, iodine, and quats are the three most common types of chemical sanitizer used in the restaurant and foodservice industry.

▶ **Chlorine:** The most commonly used sanitizer due to its low cost and effectiveness. Chlorine kills a wide range of microorganisms. However, soil can quickly inactivate chlorine solutions, and they can be corrosive to some metals when used improperly.

▶ **Iodine:** Sanitizer effective at low concentrations and not as quickly inactivated by soil as chlorine. It is somewhat corrosive to surfaces and is less effective than chlorine.

▶ **Quaternary ammonium compounds (quats):** Group of sanitizers all having the same basic chemical structure. Quats are noncorrosive to surfaces, and remain active for short periods of time after they have dried. However, quats may not kill certain types of microorganisms, and they are easily affected by detergent residue.

▶ **Material Safety Data Sheets (MSDS):** Sheets supplied by the chemical manufacturer listing the chemical and its common names, its potential physical and health hazards, information about using and handling it safely, and other important information. OSHA requires employers to store these sheets so they are accessible to employees.

▶ **Master cleaning schedule:** Detailed schedule listing all cleaning tasks in an establishment, when and how they are to be performed, and who will do them.

▶ **Integrated pest management (IPM):** Program using prevention measures to keep pests from entering an establishment and control measures to eliminate any pests that do get inside.

▶ **Pest control operator (PCO):** Licensed professional who uses safe, up-to-date methods to prevent and control pests.

▶ **Infestation:** Situation that exists when pests overrun or inhabit an establishment in large numbers.

INTRODUCTION

Sanitary facilities and equipment are basic components of a well-designed food safety system. In this section, you will find a wide range of information on designing a sanitary facility, as well as the cleaning and sanitizing practices and procedures that must be put in place to keep it that way. Keeping your establishment clean and sanitary will help prevent food from becoming contaminated. It will also keep pests—another potential source of contamination—from thriving in your establishment. This section will also explore the steps that need to be taken to implement a successful integrated pest management program in your establishment.

Sanitary Facilities and Equipment

When designing or remodeling a facility, consider how the building and equipment in each area will be kept clean and maintained in good repair. Facilities should be arranged so contact with contaminated sources—such as garbage or dirty tableware, utensils, and equipment—is unlikely to occur.

Consider the following when designing an establishment:

▶ Arrange equipment and fixtures to comply with sanitary standards.

▶ Select materials for walls, floors, and ceilings that will make cleaning these surfaces easier.

▶ Design the layout of utilities to prevent contamination and make cleaning easier.

▶ Be sure patrons and employees with disabilities have access to the building as required by the Americans with Disabilities Act (ADA).

▶ **Consult local regulations.** Many jurisdictions require approval of layout and design plans, prior to new construction or extensive remodeling. Even if local regulations do not require it, it is a good idea to have these plans reviewed.

MATERIALS FOR INTERIOR CONSTRUCTION

The most important consideration when selecting construction materials is how easy the establishment will be to clean and maintain. Sound-absorbent surfaces that also resist absorption of grease and moisture and reflect light will probably create an environment acceptable to your regulatory agency.

Flooring

Flooring materials in kitchen and service areas should meet requirements for health and safety, strength and durability, and appearance. The floor surfaces should be easy to clean, wear-resistant, slip-resistant, and nonporous. Flooring should be kept in good repair and replaced if damaged or worn. Nonabsorbent flooring should be used in food-preparation areas, walk-in refrigerators, warewashing areas, restrooms, and other areas subject to moisture, flushing, or spray cleaning.

CONSIDERATIONS FOR SPECIFIC AREAS OF THE FACILITY

Handwashing Stations

Handwashing stations must be conveniently located so employees will be encouraged to wash their hands often. They are required in food-preparation areas, service areas, warewashing areas, and restrooms. These stations must be operable, stocked, and maintained. A handwashing station must be equipped with the following items (see *Exhibit 11a*):

▶ **Hot and cold running water.** Hot and cold water should be supplied through a mixing valve or combination faucet at a temperature of at least 100°F (38°C).

▶ **Soap.** The soap can be liquid, bar, or powder. Liquid soap is generally preferred, and some local codes require it.

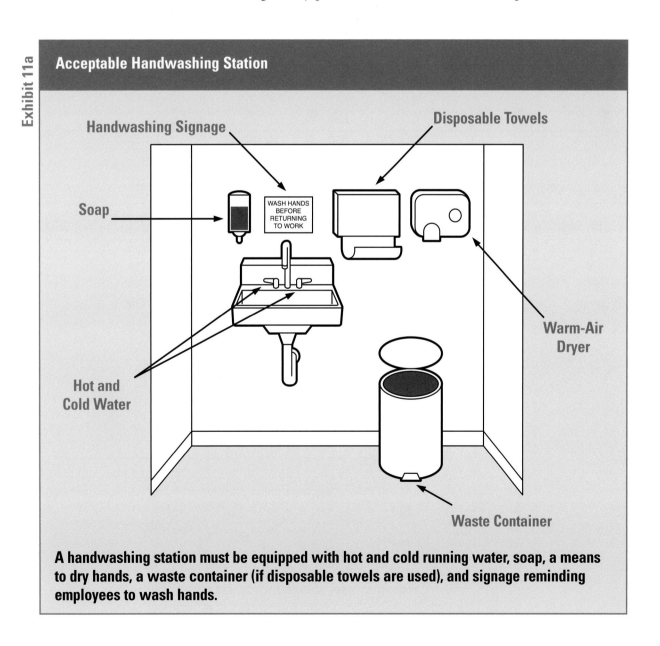

Exhibit 11a

Acceptable Handwashing Station

Handwashing Signage

Disposable Towels

Soap

WASH HANDS BEFORE RETURNING TO WORK

Warm-Air Dryer

Hot and Cold Water

Waste Container

A handwashing station must be equipped with hot and cold running water, soap, a means to dry hands, a waste container (if disposable towels are used), and signage reminding employees to wash hands.

▶ **A means to dry hands.** Most local codes require establishments to supply disposable paper towels in handwashing stations. Continuous-cloth towel systems, if allowed, should be used only if the unit is working properly and the towel rolls are checked and changed regularly. Installing at least one warm-air dryer in a handwashing station will provide an alternate method for drying hands if paper towels run out. The use of common cloth towels is not permitted because they can transmit contaminants from one person's hands to another.

▶ **Waste container.** Waste containers are required if disposable paper towels are provided.

▶ **Signage indicating employees are required to wash hands before returning to work.**

Apply Your Knowledge　　**What's Missing?**

This handwashing station is missing four items. Can you identify what's missing?

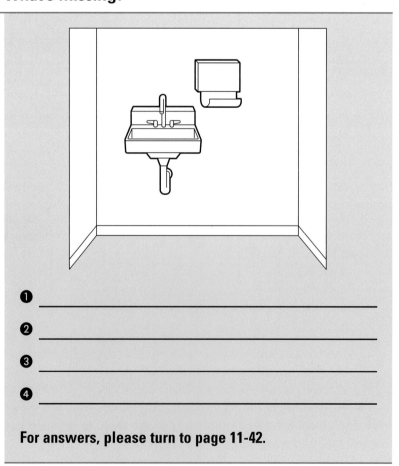

❶ _____

❷ _____

❸ _____

❹ _____

For answers, please turn to page 11-42.

SANITATION STANDARDS FOR EQUIPMENT

It is important to purchase equipment that has been designed with sanitation in mind. Food-contact surfaces must be

► safe.

► durable.

► corrosion-resistant.

► non-absorbent.

► sufficient in weight and thickness to withstand repeated warewashing.

► smooth and easy to clean.

► resistant to pitting, chipping, crazing (spider cracks), scratching, scoring, distortion, and decomposition.

Equipment surfaces that are not designed to come in contact with food, but which are exposed to splash, spillage, or other food soiling, or that require frequent cleaning, must be

► constructed of smooth, nonabsorbent, corrosion-resistant material.

► free of unnecessary ledges, projections, and crevices.

► designed and constructed to allow easy cleaning and maintenance.

The task of choosing equipment designed for sanitation has been simplified by organizations such as NSF International and Underwriters Laboratories (UL). NSF International develops and publishes standards for sanitary equipment design. The presence of the NSF mark on foodservice equipment means it has been evaluated, tested, and certified by NSF International as meeting international commercial food equipment standards. UL similarly provides sanitation classification listings for equipment found in compliance with NSF International standards. UL also lists products complying with their own published environmental and public health standards. Restaurant and foodservice managers should look for the NSF International mark or the UL EPH product mark on commercial foodservice equipment. (See the examples in *Exhibit 11b.*)

Exhibit 11b

NSF and UL EPH Marks

Look for the NSF International mark or UL EPH product marks on sanitary equipment.

Only commercial foodservice equipment should be used in establishments, since household equipment is not built to withstand heavy use. Although all equipment used in an establishment must meet standards such as those set by NSF International, certain equipment requires particular attention.

Warewashing Machines

Warewashing machines vary widely by size, style, and method of sanitizing. High-temperature machines sanitize with extremely hot water; chemical-sanitizing machines use a chemical solution. Some states require the local regulatory agency's approval before installing a chemical warewashing system.

Consider the following guidelines regarding the installation and use of warewashing machines:

▶ Water pipes to the warewashing machine should be as short as possible to prevent the loss of heat from hot water entering the machine.

▶ The machine must be raised at least six inches off the floor to permit easy cleaning underneath.

▶ Materials used in warewashing machines should be able to withstand wear, including the action of detergents and sanitizers.

▶ Information should be posted on or near the machine regarding proper water temperature, conveyer speed, water pressure, and chemical concentration.

▶ The machine's thermometer should be located so it is readable, with a scale in increments no greater than 2°F (1°C).

Clean-in-Place Equipment

Some equipment is designed to be cleaned and sanitized by having detergent solution, a hot-water rinse, and sanitizing solution pass through it. Certain soft-serve ice cream and frozen yogurt dispensers are cleaned and sanitized this way. This process should be performed daily unless otherwise indicated by the manufacturer. Instructions should be followed carefully. Cleaning and sanitizing solutions must

▶ remain within the tubes and pipes for a predetermined amount of time.

▶ reach all food-contact surfaces.

▶ not leak into the rest of the machine.

Refrigerators and Freezers

When purchasing a refrigerator or freezer, make sure it carries the NSF International mark, UL EPH product marks, or the equivalent.

INSTALLING AND MAINTAINING KITCHEN EQUIPMENT

Plan your kitchen layout so the facility and equipment is easy to clean, the risk of cross-contamination is minimized, and the time food spends in the temperature danger zone is reduced.

Installing Kitchen Equipment

Consider the following when installing kitchen equipment:

▶ Portable equipment is often easier to clean and clean around than permanently installed equipment.

Exhibit 11c

Installing Stationary Equipment

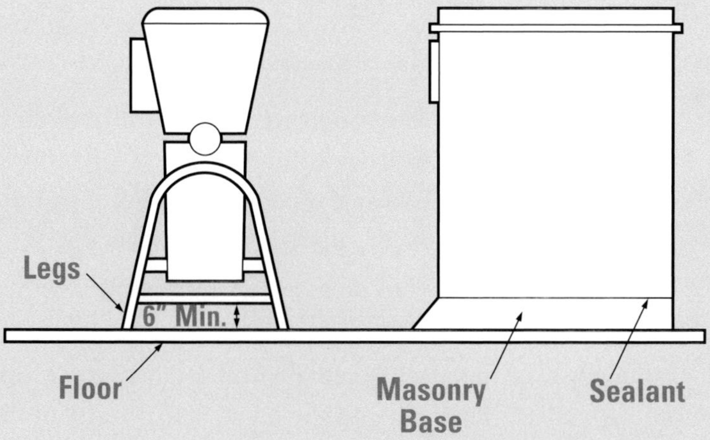

Legs

6" Min.

Floor

Masonry Base

Sealant

Stationary equipment must be mounted on legs at least six inches off the floor or it must be sealed to a masonry base.

▶ Stationary equipment must be mounted on legs at least six inches off the floor or sealed to a masonry base. (See *Exhibit 11c.*)

▶ Stationary tabletop equipment should be mounted on legs providing a minimum clearance of four inches between the base of the equipment and the tabletop. Otherwise, the equipment should be tiltable, or it should be sealed to the countertop with a nontoxic, food-grade sealant.

▶ All cracks or seams over $\frac{1}{32}''$ (0.8 mm) must be filled with a nontoxic, food-grade sealant.

Maintaining Equipment

Once equipment has been properly installed, it must receive regular maintenance. Follow the manufacturer's recommendations and make sure it is maintained by qualified personnel.

UTILITIES

Utilities used by an establishment include water and plumbing, electricity, gas, lighting, ventilation, sewage, and waste handling. There must be enough utilities to meet the cleaning needs of the establishment, and the utility must not contribute to contamination.

Water Supply

Safe water is vital in an establishment since unsafe water can carry foodborne pathogens. Water that is safe to drink is called potable water. Sources of potable water include the following:

▶ Approved public water mains

▶ Private water sources that are regularly maintained and tested

▶ Bottled drinking water

▶ Closed, portable water containers filled with potable water

▶ On-premise water storage tanks

▶ Water transport vehicles that are properly maintained

If your establishment uses a private water supply, such as a well, rather than an approved public source, you should check with your local regulatory agency for information on inspections, testing, and other requirements. Generally, nonpublic water systems should be tested at least annually and the report kept on file in the establishment.

Occasionally, an emergency occurs that causes the water supply to become unusable. Depending on the nature and severity of the problem, an establishment may wish to remain open. If the water supply is disrupted, follow these guidelines to continue serving food safely:

▶ Use bottled water.

▶ Boil water (check with the local regulatory agency).

▶ Purchase ice.

▶ Use boiled water for essential cleaning, such as pots and pans. Consider using single-use items (plates and utensils) to minimize warewashing. Keep a supply of boiled, warm water available for handwashing.

Contact the local regulatory agency if you have any questions about the safety of a particular practice.

Plumbing

Improperly installed or poorly maintained plumbing that allows the mixing of potable and nonpotable water has been implicated in foodborne-illness outbreaks. Only licensed plumbers should install and maintain plumbing systems in an establishment.

Cross-Connections

The greatest challenge to water safety comes from cross-connections. A cross-connection is a physical link through which contaminants from drains, sewers, or other waste-water sources can enter a potable water supply. A cross-connection is dangerous because it allows the possibility of backflow. Backflow is the unwanted reverse flow of contaminants through a cross-connection into a potable water system. It can occur whenever the pressure in the potable water supply drops below the pressure of the contaminated supply. A running faucet located below the flood rim of a sink or a running hose in a mop bucket (see *Exhibit 11d*) are examples of a cross-connection.

Exhibit 11d

Common Cross-Connection

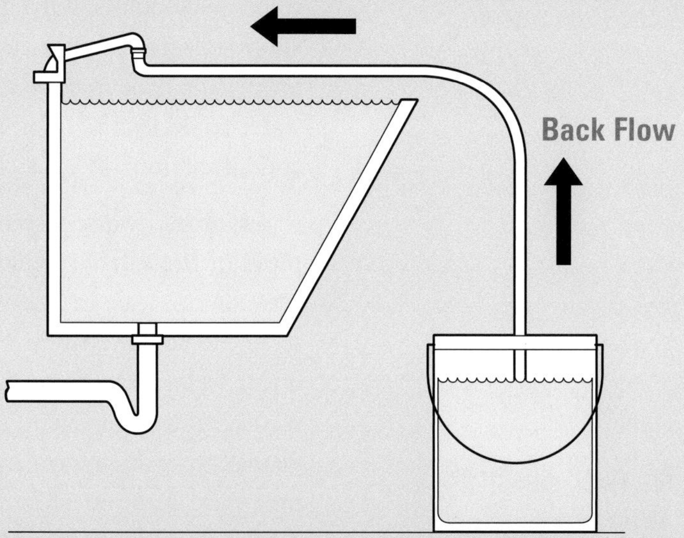

Back Flow

A hose connected to a faucet and left submerged in a mop bucket creates a dangerous cross-connection.

To prevent cross-connections, do not attach a hose to a faucet unless a backflow-prevention device, such as a vacuum breaker, is attached. Threaded faucets and connections between two piping systems must have a vacuum breaker or other approved backflow-prevention device.

The only completely reliable method for preventing backflow is creating an air gap. An air gap is an air space used to separate a water supply outlet from any potentially contaminated source. A properly designed and installed sink typically has two air gaps to prevent backflow. One is the air space between the faucet and the flood rim of the sink. The other is located between the drainpipe of the sink and the floor drain of the establishment. (See *Exhibit 11e*.)

Grease Condensation and Leaking Pipes

Grease condensation in pipes is another common problem in plumbing systems. Grease traps are often installed to prevent a buildup from creating a drain blockage. If used, grease traps must be easily accessible, installed by a licensed plumber, and cleaned periodically according to manufacturers' recommendations. If the traps are not cleaned, or not cleaned properly, a backup of waste water could lead to odor and contamination.

Overhead waste-water pipes or fire-safety sprinkler systems can leak and become a source of contamination. Even overhead lines carrying potable water can be a problem, since water can condense on the pipes and drip onto food. All piping should be serviced immediately when leaks occur.

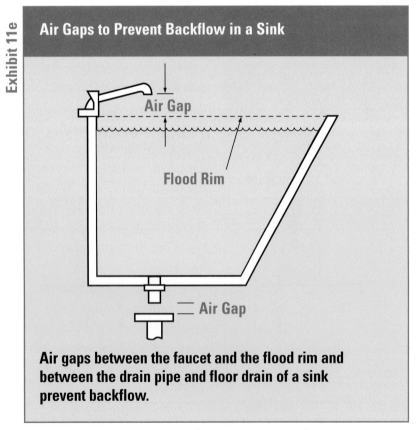

Exhibit 11e

Air Gaps to Prevent Backflow in a Sink

Air Gap

Flood Rim

Air Gap

Air gaps between the faucet and the flood rim and between the drain pipe and floor drain of a sink prevent backflow.

Sewage

The facility must have adequate drainage to handle all waste water. Areas subject to heavy water exposure should have floor drains.

Sewage and waste water are contaminated with pathogens, soils, and chemicals. It is absolutely essential to prevent them from contaminating food or food-contact surfaces.

A backup of raw sewage is cause for immediate closure of the establishment, correction of the problem, and thorough cleaning.

Lighting

Building and health codes usually set minimum acceptable levels of lighting, typically based on the foot-candle—a unit of illumination one foot from a uniform source of light. Other units of measurement for light include lumens, luxes, and luminaires. Good lighting generally results in improved employee work habits, easier and more effective cleaning, and a safer work environment. Lighting intensity requirements are different for various areas of the establishment. (See *Exhibit 11f* on the next page.)

You should also consider the following regarding lighting:

▶ **Position overhead or ceiling lights above workstations so employees do not cast shadows on the work surface.** Using fluorescent lights helps minimize such shadows.

▶ **Use shatter-resistant light bulbs and protective covers made of metal mesh or plastic.** This will prevent broken glass from contaminating food or food-contact surfaces.

▶ **Provide shields for heat lamps.**

Ventilation

Ventilation helps maintain an establishment's indoor air quality by removing odors, gases, grease, dirt, and mold that can cause contamination. If ventilation is adequate, there will be little or no build-up of grease and condensation on walls and ceilings. Ventilation must be designed so hoods, fans, guards, and duct work do not drip onto food or equipment. Hood filters or grease extractors must be tight-fitting and easy to remove, and should be

Exhibit 11f

Minimum Lighting Intensity Requirements for Different Areas of the Establishment	
Minimum Lighting Intensity	**Area**
50 foot-candles (540 lux)	▶ Food-preparation areas
20 foot-candles (220 lux)	▶ Handwashing or warewashing areas ▶ Buffets and salad bars ▶ Displays for produce or packaged food ▶ Utensil-storage areas ▶ Wait stations ▶ Restrooms ▶ Inside some pieces of equipment (e.g., reach-in refrigerators)
10 foot-candles (110 lux)	▶ Inside walk-in refrigerators and freezer units ▶ Dry-storage areas ▶ Dining rooms (for cleaning)

cleaned on a regular basis. Thorough cleaning of the hood and duct work should also be done periodically by a professional company. It is the establishment's responsibility to see that the ventilation system meets local regulations.

Garbage Disposal

Garbage is wet waste matter, usually containing food, that cannot be recycled. It attracts pests and can contaminate food, equipment, and utensils.

To control hazards that garbage can pose, consider the following:

▶ Garbage should be removed from food-preparation areas as quickly as possible to prevent odors, pests, and possible contamination. Do not carry garbage above or across food-preparation areas.

▶ Garbage containers must be leak proof, waterproof, and pest proof, and have tight-fitting lids. Typically, they should be made of galvanized metal or an approved plastic, and they should be easy to clean.

▶ Plastic bags and wet-strength paper bags may be used to line garbage containers.

▶ Garbage containers should be cleaned frequently and thoroughly, both inside and out. This will help keep odors and pests to a minimum. Areas used for cleaning garbage containers should not be located near areas used for food preparation or storage.

▶ Outdoor trash receptacles should be kept covered at all times (with their drain plugs in place).

Apply Your Knowledge

Identify the minimum lighting intensity requirement for each area and write the letter in the space provided.

How Bright Should It Be?

___ ❶ Dry-storage areas

___ ❷ Food-preparation areas

___ ❸ Handwashing areas

___ ❹ Walk-in refrigerators

___ ❺ Warewashing areas

    A. 50 foot-candles (540 lux)

    B. 20 foot-candles (220 lux)

    C. 10 foot-candles (110 lux)

For answers, please turn to page 11-42.

Cleaning and Sanitizing

Once you have designed a sanitary facility, it is important to keep it that way. If you do not keep the facility and equipment clean and sanitary, food can easily become contaminated. To prevent this, it is important to understand the difference between cleaning and sanitizing. *Cleaning* is the process of removing food and other types of soil from a surface, such as a countertop or plate. *Sanitizing* is the process of reducing the number of microorganisms on a clean surface to safe levels. To be effective, cleaning and sanitizing must be a two-step process. Surfaces must *first* be cleaned and rinsed *before* being sanitized.

Everything in your operation must be kept clean; however, any surface that comes in contact with food must be cleaned *and* sanitized. All food-contact surfaces must be washed, rinsed, and sanitized

▶ after each use.

▶ anytime you begin working with another type of food.

▶ anytime you are interrupted during a task and the tools or items you have been working with may have been contaminated.

▶ at four-hour intervals, if the items are in constant use.

Apply Your Knowledge

To Sanitize or Not to Sanitize

Place an **X** next to each situation that requires the foodhandler to clean and sanitize the item he or she is using to prepare food.

___ ❶ Jorge has used the same knife to shuck oysters for two hours.

___ ❷ Bill finishes deboning chicken and wants to use the same cutting board to filet fish.

___ ❸ Mary returns to the slicer to continue slicing ham after being called away to help with the lunch rush.

___ ❹ Nora has just used the last of the pooled egg mixture from a bowl and wants to use the bowl to make a new batch.

___ ❺ Maria, a deli employee, has been slicing cheese on the same slicer from 8:00 A.M. to 12:00 P.M.

For answers, please turn to page 11-42.

CLEANING AGENTS

Cleaning agents are chemical compounds that remove food, soil, rust, stains, minerals, and other deposits. They must be stable, noncorrosive, and safe for employees use. Ask your supplier to help you select the cleaning agents that best meet your needs. When using cleaning agents, follow manufacturers' instructions carefully, since they can be ineffective and even dangerous if misused.

SANITIZING

There are two methods used to sanitize surfaces—heat sanitizing and chemical sanitizing. Which you use depends on the application.

Heat Sanitizing

The higher the heat, the shorter the time required to kill microorganisms. The most common way to heat-sanitize tableware, utensils, or equipment is to immerse or spray them with hot water. Use a thermometer to check water temperature when heat-sanitizing by immersion. To check the water temperature in a high-temperature warewashing machine, attach a temperature-sensitive label or tape, or a high-temperature probe to items being run through the machine.

Chemical Sanitizing

Chemical sanitizers are regulated by state and federal environmental protection agencies (EPAs). The three most common types are chlorine, iodine, and quaternary ammonium compounds (quats). **Refer to your local or state regulatory agency for recommendations on selecting a sanitizer.** For a list of approved sanitizers, check the Code of Federal Regulations (CFR) 21CFR178.1010—"Sanitizing Solutions."

Chemical sanitizing is done in two ways: either by immersing a clean object in a specific concentration of sanitizing solution for a required period of time, or by rinsing, swabbing, or spraying the object with a specific concentration of sanitizing solution.

In some instances, detergent-sanitizer blends may be used to sanitize surfaces, but items still must be cleaned and rinsed first. Scented or oxygen bleaches are not acceptable as sanitizers for food-contact surfaces. Household bleaches are acceptable only if the labels indicate they are EPA registered.

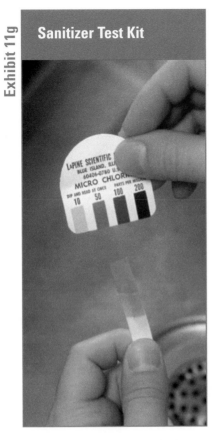

Sanitizer Test Kit

Use a test kit to check the concentration of a sanitizing solution.

Factors Influencing the Effectiveness of Sanitizers

Different factors influence the effectiveness of chemical sanitizers. The most critical include:

► **Concentration.** Chemical sanitizers are mixed with water until the proper concentration—ratio of sanitizer to water—is reached. Concentration is measured using a sanitizer test kit and is expressed in parts per million (ppm). The test kit should be designed for the sanitizer you are using and is usually available from the manufacturer or your supplier. (See *Exhibit 11g.*) The concentration of a sanitizing solution must be checked frequently since the sanitizer is depleted during use. It can also become bound up by hard water, food particles, or detergent that is not adequately rinsed from a surface. A sanitizing solution must be changed when it is visibly dirty, or when its concentration has dropped below the required level.

► **Temperature.** Generally, sanitizers work best at temperatures between 55°F and 120°F (13°C and 49°C).

► **Contact time.** For a sanitizing solution to kill microorganisms, it must make contact with the object for a specific amount of time.

See *Exhibit 11h* for the factors that influence the effectiveness of chlorine, iodine, and quats.

MACHINE WAREWASHING

Most tableware, utensils, and even pots and pans can be cleaned and sanitized in a warewashing machine. Warewashing machines sanitize by using either hot water or a chemical sanitizing solution.

High-Temperature Machines

► These machines rely on hot water to clean and sanitize.

► The temperature of the final sanitizing rinse must be at least 180°F (82°C). For stationary rack, single temperature machines, it must be at least 165°F (74°C).

► Make sure your warewasher has a built-in thermometer to measure the temperature of water at the manifold, where it sprays into the tank.

Exhibit 11h

General Guidelines for Using Chlorine, Iodine, and Quats							
	Chlorine					Iodine	Quats
Temperature	120°F (49°C)	120°F (49°C)	100°F (38°C)	75°F (24°C)	55°F (13°C)	75°F (24°C)	75°F (24°C)
Concentration	25 ppm	25 ppm	50 ppm	50 ppm	100 ppm	12.5–25 ppm	Up to 200 ppm, or as per manufacturer's recommendations
pH	10	8	10	8	8–10	≤5.0	As per manufacturer's recommendations
Contact Time	10 sec	10 sec	7 sec	7 sec	10 sec	30 sec	30 sec

Chemical-Sanitizing Machines

▶ These machines use chemicals to sanitize.

▶ Chemical sanitizing machines often wash at much lower temperatures, but not lower than 120°F (49°C).

▶ Rinse-water temperature in these machines should be between 75°F and 120°F (24°C and 49°C) for the sanitizer to be effective.

Warewashing Machine Operation

All warewashing machines should be operated according to manufacturers' instructions. No matter what type of machine you use, however, there are some general procedures to follow to clean and sanitize tableware, utensils, and related items.

▶ **Check the machine for cleanliness at least once a day, cleaning it as often as needed.** Fill tanks with clean water. Clear detergent trays and spray nozzles of food and foreign objects. Use an acid cleaner on the machine whenever necessary to remove mineral deposits caused by hard water.

▶ **Make sure detergent and sanitizer dispensers are properly filled.**

▶ **Scrape, rinse, or soak items before washing.** Presoak items with dried-on food.

▶ **Load warewasher racks correctly and use racks designed for the items being washed.** Make sure all surfaces are exposed to the spray action. Never overload racks.

▶ **Check temperatures and pressure.** Follow manufacturers' recommendations.

▶ **Check each rack as it comes out of the machine for soiled items.** Run dirty items through again until they are clean. Most items will need only one pass if the water temperature is correct and proper procedures are followed.

▶ **Air-dry all items.** Towels can recontaminate items.

▶ **Keep your warewashing machine in good repair.**

CLEANING AND SANITIZING IN A THREE-COMPARTMENT SINK

Establishments that do not have a warewashing machine may use a three-compartment sink to wash items (some local regulatory agencies allow the use of two-compartment sinks; others require four-compartment sinks). These sinks may also be used to wash larger items. A properly set-up warewashing station includes:

▶ An area for rinsing away food or for scraping food into garbage containers

▶ Drain boards to hold both soiled and clean items

▶ A thermometer to measure water temperature

▶ A clock with a second hand, allowing employees to time how long items have been immersed in the sanitizing sink

Before cleaning and sanitizing items in a three-compartment sink, each sink and all work surfaces must be cleaned and sanitized. Follow the steps listed below when cleaning and sanitizing tableware, utensils, and equipment. (See *Exhibit 11i.*)

❶ **Rinse, scrape, or soak all items before washing.**

❷ **Wash items in the first sink in a detergent solution at least 110°F (43°C).** Use a brush, cloth, or nylon scrub pad to loosen the remaining soil. Replace the detergent solution when the suds are gone or the water is dirty.

❸ **Immerse or spray-rinse items in the second sink using water at least 110°F (43°C).** Remove all traces of food and detergent. If using the immersion method, replace the rinse water when it becomes cloudy or dirty.

❹ **Immerse items in the third sink in hot water or a chemical-sanitizing solution.** If hot-water immersion is used, the water must be at least 171°F (77°C)—some health codes require a temperature of 180°F (82°C)—and the items must be immersed for thirty seconds. If chemical sanitizing is used, the sanitizer must be mixed at the proper concentration and the water temperature must be correct. Check the concentration of the sanitizing solution at regular intervals with a test kit.

❺ **Air-dry all items.**

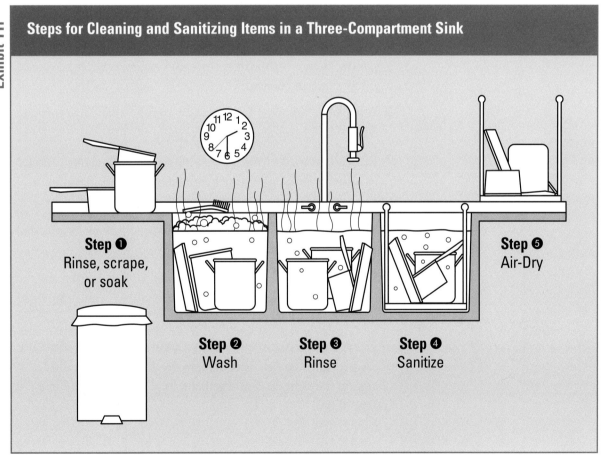

Exhibit 11i

Steps for Cleaning and Sanitizing Items in a Three-Compartment Sink

Step ❶
Rinse, scrape, or soak

Step ❷
Wash

Step ❸
Rinse

Step ❹
Sanitize

Step ❺
Air-Dry

CLEANING AND SANITIZING EQUIPMENT

Because equipment must be kept clean and food-contact surfaces must be cleaned and sanitized, employees should be taught how to clean each type of equipment properly.

Clean-in-Place Equipment

Some pieces of equipment, such as soft-serve yogurt machines, are designed to have cleaning and sanitizing solutions pumped through them. Since many of them hold and dispense potentially hazardous food, they must be cleaned and sanitized every day unless otherwise indicated by the manufacturer.

Stationary Equipment

Equipment manufacturers will usually provide cleaning instructions. In general, follow these steps:

❶ Turn off and unplug equipment before cleaning.

❷ Remove food and soil from under and around the equipment.

❸ Remove detachable parts and manually wash, rinse, and sanitize them, or run them through a warewasher, if permitted. Allow them to air-dry.

❹ Wash and rinse fixed food contact surfaces, then wipe or spray them with chemical sanitizing solution.

❺ Keep cloths used for food-contact and nonfood-contact surfaces in separate, properly marked containers of sanitizing solution.

❻ Air-dry all parts, then reassemble according to directions. Tighten all parts and guards. Test equipment at recommended settings, then turn it off.

❼ Resanitize food-contact surfaces handled when putting the unit back together by wiping with a cloth that has been submerged in sanitizing solution.

In some cases, you may be able to spray-clean fixed equipment. Check with the manufacturer. If allowed, spray each part with solution in the right concentration and let it sit for the recommended amount of time.

CLEANING NONFOOD-CONTACT SURFACES

Nonfood-contact surfaces such as floors, walls, ceilings, equipment exteriors, and restrooms must be cleaned regularly to prevent the accumulation of dust, dirt, food residue, and other debris.

TOOLS FOR CLEANING

Cleaning is easier when you have the right tools at hand. Keep tools used for cleaning separate from those used for sanitizing, and tools used for food-contact surfaces separate from those used for nonfood-contact surfaces. Using color-coded tools is one way to accomplish this. Always use a separate set for the restroom.

Brushes

Brushes apply more effective pressure than wiping cloths, and the bristles loosen soil more easily. Worn brushes will not clean effectively and can be a source of contamination.

Brushes come in different shapes and sizes for each task. Lacquered wood or plastic brushes with synthetic bristles are preferred. They do not absorb moisture, are nonabrasive, and last longer. Use the right brush for the job.

Scouring Pads

Steel wool and other abrasives are sometimes used to clean heavily soiled pots and pans, equipment, or floors. However, metal scouring pads can break apart and leave residue on surfaces, which can later contaminate food. Nylon scouring pads provide an alternative.

Mops and Brooms

Keep both light and heavy-duty mops and brooms on hand. Mop heads can be all cotton or synthetic blends. It makes sense to have a bucket and wringer for both the front and back of the facility. Both vertical and push-type brooms will be useful.

STORING UTENSILS, TABLEWARE, AND EQUIPMENT

Once utensils, tableware, and equipment are clean and sanitary, store them so they stay that way. It is equally important to ensure that cleaning tools and supplies are stored properly.

Tableware and Equipment

▶ Store tableware and utensils at least six inches off the floor. Keep them covered or otherwise protected from dirt and condensation.

▶ Clean and sanitize drawers and shelves before clean items are stored.

▶ Clean and sanitize trays and carts used to carry clean tableware and utensils. Do this daily or as often as necessary.

▶ Store glasses and cups upside down. Store flatware and utensils with handles up so employees can pick them up without touching food-contact surfaces.

▶ Keep the food-contact surfaces of clean-in-place equipment covered until ready for use.

Cleaning Tools and Supplies

Cleaning tools and supplies should be cleaned and sanitized before being put away. Tools and chemicals should be stored in a locked area away from food and food-preparation areas. The area should be well lighted so employees can identify chemicals easily. It should also be equipped with hooks for hanging mops, brooms, and other cleaning tools, a utility sink for filling buckets and cleaning tools, and a floor drain. (See *Exhibit 11j.*) Never use handwashing sinks, food-preparations sinks, or warewashing sinks to clean mops, brushes, or other tools. When storing tools and supplies, consider the following:

▶ Air-dry wiping cloths overnight.

▶ Hang mops, brooms, and brushes on hooks to air-dry.

▶ Clean, rinse, and sanitize buckets. Let them air-dry, and store them with other tools.

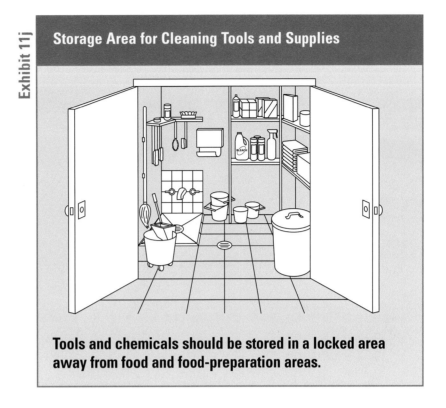

Storage Area for Cleaning Tools and Supplies

Tools and chemicals should be stored in a locked area away from food and food-preparation areas.

USING HAZARDOUS MATERIALS

To reduce your risk, you should only purchase chemicals that are approved for use in a restaurant or foodservice establishment. As previously mentioned, they should be stored in their original containers away from food and food-preparation areas. If chemicals are transferred to a new container, the label on that container must include the following information:

▶ Chemical name

▶ Manufacturer's name and address

▶ Potential hazards of the chemical

When disposing of chemicals, follow the instructions on the label and any local regulations that apply.

OSHA requires chemical manufacturers and suppliers to provide Material Safety Data Sheets (MSDS) for each hazardous chemical at your establishment. These sheets are sent periodically with shipments or can be requested by the establishment. MSDS are part of employees' right to know about the hazardous

Exhibit 11k

Sample Material Safety Data Sheets

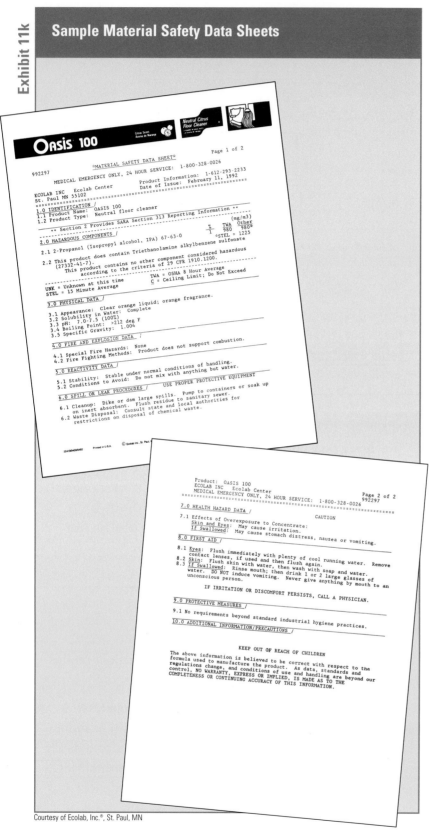

Courtesy of Ecolab, Inc.®, St. Paul, MN

chemicals they work with and must be kept in a location accessible to all employees while on the job. MSDS contain the following information about the chemical (see *Exhibit 11k*):

▶ Information about safe use and handling

▶ Physical, health, fire, and reactivity hazards

▶ Precautions

▶ Appropriate personal protective equipment (PPE) to wear when using the chemical

▶ First-aid information and steps to take in an emergency

▶ Manufacturer's name, address, and phone number

▶ Preparation date of MSDS

▶ Hazardous ingredients and identity information

IMPLEMENTING A CLEANING PROGRAM

A clean and sanitary establishment is a prerequisite for an effective food safety management system. It also takes commitment from management and the involvement of employees.

Create a Master Cleaning Schedule

Take information gathered while identifying your cleaning needs and develop a master cleaning schedule. The schedule should include the following:

▶ **What should be cleaned.** Arrange the schedule in a logical way so nothing is left out. List all cleaning jobs in one area, or list jobs in the order they should be performed.

▶ **Who should clean it.** Assign each task to a specific individual.

▶ **When it should be cleaned.** Employees should clean as they go *and* clean and sanitize at the end of their shifts. Schedule major cleaning when food will not be contaminated or service interrupted—usually after closing. Schedule work shifts to allow enough time.

▶ **How it should be cleaned.** Provide clearly written procedures for cleaning. Lead employees through the process step by step. Always follow manufacturers' instructions when cleaning equipment. Specify cleaning tools and chemicals by name. Post cleaning instructions near the item to be cleaned.

Monitor the Program

▶ Supervise daily cleaning routines.

▶ Monitor the daily completion of all cleaning tasks against the master cleaning schedule.

▶ Review the master schedule every time there is a change in menu, procedures, or equipment.

▶ Request employee input on the program during staff meetings.

▶ Conduct spot inspections.

Apply Your Knowledge	**What's Wrong with This Picture?**

There are several things wrong with this three-compartment sink. Identify as many as you can in the space provided.

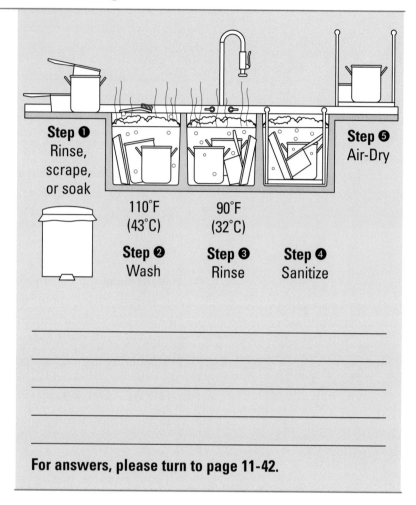

Step ❶ Rinse, scrape, or soak

110°F (43°C)

90°F (32°C)

Step ❷ Wash

Step ❸ Rinse

Step ❹ Sanitize

Step ❺ Air-Dry

For answers, please turn to page 11-42.

Integrated Pest Management (IPM)

Keeping your establishment clean and sanitary will help prevent food from becoming contaminated. It will also keep pests—another potential source of contamination—from thriving in your establishment, since a clean and sanitary establishment offers them little in the way of food and shelter.

Pests such as insects and rodents can pose serious problems for establishments. Not only are they unsightly to customers, they also damage food, supplies, and facilities. However, the greatest danger from pests comes from their ability to spread diseases, including foodborne illnesses.

THE INTEGRATED PEST MANAGEMENT PROGRAM

Once pests have come into a facility in large numbers—an infestation—they can be very difficult to eliminate. Developing and implementing an integrated pest management (IPM) program is the key. An IPM program uses *prevention* measures to keep pests from entering the establishment and *control* measures to eliminate any pests that do get inside.

For your IPM program to be successful, it is best to work closely with a licensed pest control operator (PCO). These professionals use safe, up-to-date methods to prevent and control pests. Prevention is critical in pest control. If you wait until there is evidence of pests in your establishment, they may already be there in large numbers.

There are three basic rules of an IPM program:

❶ Deny pests access to the establishment.

❷ Deny pests food, water, and a hiding or nesting place.

❸ Work with a licensed PCO to eliminate pests that do enter.

DENYING PESTS ACCESS TO THE ESTABLISHMENT

Pests can enter an establishment in one of two ways. They either are brought inside with deliveries, or they enter through openings in the building itself. To prevent pests from entering your establishment, pay particular attention to the following areas.

Deliveries

▶ Use reputable suppliers.

▶ Check all deliveries before they enter your establishment.

▶ Refuse shipments in which you find pests or signs of infestation, such as egg cases and body parts (legs, wings, etc.).

Doors, Windows, and Vents

▶ Screen all windows and vents with at least sixteen mesh per square inch screening.

▶ Install self-closing devices or door sweeps on all doors.

▶ Install air curtains (also called air doors or fly fans) above or alongside doors. These devices blow a steady stream of air across the entryway, creating an air shield around doors left open.

▶ Keep all exterior openings closed tightly.

Pipes

Mice, rats, and insects use pipes as highways through a facility.

▶ Use concrete to fill holes or sheet metal to cover openings around pipes. (See *Exhibit 11l.*)

▶ Install screens over ventilation pipes and ducts on the roof.

▶ Cover floor drains with hinged grates to keep rodents out.

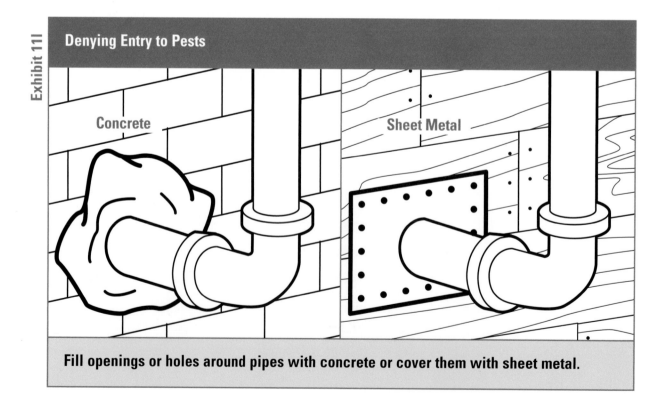

Exhibit 11l

Denying Entry to Pests

Concrete   Sheet Metal

Fill openings or holes around pipes with concrete or cover them with sheet metal.

Floors and Walls

Rodents often burrow into buildings through decaying masonry or cracks in building foundations. They move through floors and walls the same way. Mice can squeeze through a hole the size of a dime, rats through holes the size of a quarter.

▶ Seal all cracks in floors and walls. Use a permanent sealant recommended by your PCO or local health department.

▶ Properly seal spaces or cracks where fixed equipment is fitted to the floor.

DENYING PESTS FOOD AND SHELTER

Pests are usually attracted to damp, dark, and dirty places. A clean and sanitary establishment offers them little in the way of food and shelter. The stray pest that might get in cannot thrive or multiply in a clean kitchen. Besides adhering to your master cleaning schedule, follow these additional guidelines.

▶ **Dispose of garbage quickly and correctly.** Keep garbage containers clean, in good condition, and tightly covered in all areas (indoor and outdoor). Clean up spills around garbage containers immediately. Wash and rinse containers regularly.

▶ **Store recyclables in clean, pest-proof containers as far away from your building as local regulations allow.**

▶ **Store all food and supplies properly and as quickly as possible.**

  ▶ Keep all food and supplies away from walls and at least six inches off the floor.

  ▶ Keep humidity in dry storerooms as low as possible. Low humidity helps prevent roach eggs from hatching.

  ▶ Refrigerate food such as powdered milk, cocoa, and nuts after opening. Most insects that might be attracted to this food become inactive at temperatures below 41°F (5°C).

  ▶ Rotate products, so pests do not have time to settle into them and breed.

Common Cockroaches Found in Restaurants and Foodservice Operations

American

German

Brown-banded

Oriental

Courtesy of Orkin Commercial Services

▶ **Clean the facility thoroughly.** Careful cleaning eliminates the food supply, destroys insect eggs, and reduces the number of places pests can safely take shelter.

▶ **Deny food and shelter in outdoor dining areas.** This includes pulling weeds, removing standing water, picking up litter, and removing dirty dishes and uneaten food from tables.

IDENTIFYING PESTS

Pests may still get into your establishment even if you take careful preventive measures. Learn how to spot signs of pests and identify what kind they are. If possible, record the time, date, and location of any pest sighting (or evidence of pests) and report this to your PCO. Early detection gives your PCO a chance to start treatment as soon as possible.

Cockroaches

Roaches often carry disease-causing microorganisms such as *Salmonella* spp., fungi, parasite eggs, and viruses. Most live and breed in dark, warm, moist, and hard-to-clean places. If you see a cockroach in daylight, you may have a major infestation since only the weakest roaches come out at that time. If you suspect you have a roach problem, check for these signs:

▶ **A strong oily odor**

▶ **Droppings (feces), which look like grains of black pepper**

▶ **Capsule-shaped egg cases that are brown, dark red, or black and may appear leathery, smooth, or shiny**

Glue Traps

You may have problems with more than one type of roach. Glue traps—containers with sticky glue on the bottom—should be used to find out what type of roaches might be present. When using glue traps:

▶ **Work with your PCO to place them where roaches typically can be found.** If possible, place them on the floor in the corner where two walls meet.

▶ **Check them after twenty-four hours and show them to your PCO.** The type and stage (nymph or adult) of roaches present will determine the type of treatment needed.

Common Types of Rodents

Roof Rat

Common House Mouse

Norway Rat

Courtesy of Orkin Commercial Services

Rodents

Rodents are a serious health hazard. They eat and ruin food, damage property, and can spread disease. A building can be infested with both rats and mice at the same time. Look for the following signs.

▶ **Signs of gnawing.** Rats and mice gnaw to reach food and to wear down their teeth, which grow continuously.

▶ **Droppings.** Fresh droppings are shiny and black. Older droppings are gray.

▶ **Tracks.** Check dusty surfaces by shining a light across them at a low angle.

▶ **Nesting materials.** Mice use scraps of paper, cloth, hair, and other soft materials to build nests.

▶ **Holes.** Rats nest in burrows, usually in dirt, rock piles, or along foundations.

WORKING WITH A PEST CONTROL OPERATOR

Few pest problems are solved simply by spraying pesticides—chemical agents used to destroy pests. Although you can take most preventive measures yourself, most control measures should be carried out only by professionals. Employ a licensed, certified PCO to handle pest control. Working as a team, you and the PCO can prevent and/or eliminate pests and keep them from coming back.

USING AND STORING PESTICIDES

To minimize the hazard to people, have your PCO apply pesticides only when you are closed for business and employees are not on-site. Follow these guidelines when pesticides will be applied:

▶ Prepare the area to be sprayed by removing all food and food-contact surfaces.

▶ Cover equipment and food-contact surfaces that cannot be moved.

▶ Wash, rinse, and sanitize food-contact surfaces after the area has been sprayed.

Your PCO should store and dispose of all pesticides used in your facility. If they are stored on the premises, follow these guidelines:

▶ **Keep pesticides in their original containers.**

▶ **Store pesticides in locked cabinets away from areas where food is stored and prepared.**

▶ **Check local regulations before disposing of pesticides.** Many are considered hazardous waste. Dispose of empty containers according to manufacturers' directions and local regulations.

▶ **Keep a copy of the corresponding MSDS on the premises.**

Apply Your Knowledge	**Who Am I?**

In the space provided, place a **C** next to the characteristics of cockroaches, and an **R** next to the characteristics of rodents.

___ ❶ I nest in scraps of paper, cloth, and hair.

___ ❷ I produce a strong oily odor.

___ ❸ Glue traps are often used to find out who I am.

___ ❹ My droppings are shiny black or gray.

___ ❺ I can burrow into the establishment along the foundation.

___ ❻ My droppings look like grains of black pepper.

___ ❼ If you see me in the daytime, I am present in large numbers.

For answers, please turn to page 11-42.

SUMMARY

An establishment that is difficult to clean will not be cleaned well. Sanitation efforts will be more effective if a facility is designed and equipped with ease of cleaning in mind. Purchase equipment with food-contact surfaces that are corrosion-resistant, nonabsorbent, smooth, and resist pitting and scratching. Equipment should be installed so that both the equipment and the surrounding area can be cleaned easily. Stationary equipment must be mounted on legs at least six inches off the floor, or it must be sealed to a masonry base. Stationary tabletop equipment should be mounted on legs with a clearance of four inches between the equipment bottom and the tabletop, or it should be sealed to the tabletop. Handwashing stations must be conveniently located, operable, and fully stocked and maintained. They are required in food-preparation areas, service areas, warewashing areas, and restrooms.

Utilities must be designed to meet the establishment's cleaning needs, and they must not contribute to contamination. Plumbing should be installed and maintained by licensed plumbers. Vacuum breakers and air gaps should be used to prevent the backflow of contaminants into the potable water supply. It is essential to prevent waste water from contaminating food and food-contact surfaces. A back-up of raw sewage is cause for immediate closure, correction of the problem, and thorough cleaning. Good lighting in an establishment will result in more effective cleaning and a safer work environment. Follow the lighting intensity requirements for each area of the establishment. Ventilation must be designed so hoods, fans, guards, and ductwork do not drip onto food or equipment. Hood filters and grease extractors must be cleaned regularly. Garbage containers must be leak-proof, waterproof, pest-proof, easy-to-clean, and durable. They must have tight-fitting lids, be kept covered when not in use, and be cleaned frequently. Garbage should be removed from food-preparation areas as soon as possible.

Cleaning is the process of removing food and other types of soil from a surface. Sanitizing is the process of reducing the number of harmful microorganisms on a clean surface to safe levels. You must clean and rinse a surface before it can be sanitized effectively. Surfaces can be sanitized with hot water or with a chemical-sanitizing solution. All surfaces should be cleaned on a regular basis. Food-contact surfaces must be cleaned and then sanitized after every use, anytime a task is interrupted, and at four-hour intervals if items are in constant use.

Follow manufacturers' instructions when using warewashing machines. Check the temperature and pressure of the wash and rinse cycles daily. Items can be cleaned and sanitized in a three-compartment sink, or, if the items are stationary, by cleaning and then spraying them with a sanitizing solution. Items cleaned and sanitized in a three-compartment sink should be presoaked or scraped clean, washed in a detergent solution, rinsed in clean water, and sanitized in either hot water or immersed in a sanitizing solution for a predetermined amount of time. All items should then be air-dried.

Cleaning tools and supplies should be stored in a well-lighted, locked room separate from areas where food is prepared or stored. Chemicals should be labeled clearly, and MSDS should be on hand for each chemical. Create a master cleaning schedule listing all cleaning tasks and identifying when and how they are to be performed. Assign responsibility for each task by job title. Monitor the cleaning program to keep it effective.

Pests are a threat to establishments because they can carry and spread a variety of diseases. Once they have infested a facility, it can be very difficult to eliminate them. Developing and implementing an integrated pest management (IPM) program is the key. An IPM program uses prevention measures to keep pests from entering the establishment and control measures to eliminate any pests that do get inside. For your IPM program to be successful, it is best to work closely with a PCO.

Understanding pests is the key to controlling them. Roaches live and breed in dark, warm, and moist places. Check for a strong oily odor, droppings—which look like grains of black pepper—and egg cases. Rodents also are a serious health hazard. Look for droppings, signs of gnawing, tracks, nesting materials, and holes. Although you can take most prevention measures yourself, most control measures should be carried out by a licensed, certified PCO. Pesticides are hazardous materials. Anytime they are used or stored on the premises, you should have corresponding MSDS. To minimize the hazard to people, have your PCO use pesticides only when you are closed for business and your employees are not on-site. Your PCO should store and dispose of all pesticides used in your facility. If they are stored on the premises, they should be kept in their original containers and stored in locked cabinets away from food-storage and food-preparation areas.

Apply Your Knowledge

Use these questions to review the concepts presented in this section.

Multiple-Choice Study Questions

1. Generally, establishments that use a private water source, such as a well, must have it tested at least
   A. once a year.
   B. every two years.
   C. every three years.
   D. every five years.

2. Which of the following will *not* prevent backflow?
   A. An air gap between the sink drain pipe and the floor drain
   B. The air space between the faucet and the flood rim of a sink
   C. A vacuum breaker
   D. A cross-connection

3. When mounting tabletop equipment on legs, the clearance between the base of the equipment and the tabletop must be at least
   A. one inch.
   B. two inches.
   C. four inches.
   D. six inches.

4. Equipment food-contact surfaces must meet all of the following conditions *except*
   A. they must be corrosion-resistant.
   B. they must be absorbent.
   C. they must be smooth.
   D. they must be resistant to pitting.

5. The lighting intensity in a dry-storage area should be at least
   A. fifty foot-candles (540 lux).
   B. twenty foot-candles (220 lux).
   C. ten foot-candles (110 lux).
   D. five foot-candles (54 lux).

6. Which of the following items need to be both cleaned and sanitized?
   A. Floors
   B. Walls
   C. Cutting boards
   D. Ceilings

Continued on next page...

Apply Your Knowledge   Multiple-Choice Study Questions *continued*

7. You want to make a spray solution for use in sanitizing food-contact surfaces in the establishment. What should you do to ensure that you have made a proper sanitizing solution?
    A. Compare the color of the solution to another solution of known strength.
    B. Try out the solution on a food-contact surface.
    C. Test the solution with a sanitizer test kit.
    D. Use very hot water when making the solution.

8. What is the proper procedure for sanitizing a table that has been used to prepare food?
    A. Spray it with a strong sanitizing solution, then wipe it dry.
    B. Wash it with a detergent, rinse it, then wipe it with a sanitizing solution.
    C. Wash it with a detergent, then wipe it dry.
    D. Wipe it with a dry cloth, then wipe it with a sanitizing solution.

9. If food-contact surfaces are in constant use, they must be cleaned and sanitized at
    A. four-hour intervals.
    B. five-hour intervals.
    C. six-hour intervals.
    D. eight-hour intervals.

10. Which of the following is an improper method for storing clean and sanitized tableware and equipment?
    A. Storing glasses and cups upside down
    B. Storing tableware six inches off the floor
    C. Storing flatware in containers with the handles down
    D. Storing utensils in a covered container until needed

11. Cockroaches typically are found in places that are
    A. cold, dry, and light.   C. warm, moist, and dark.
    B. cold, moist, and dark.   D. warm, dry, and light.

Continued on next page...

12. Which of the following is a sign you might have a problem with rodents?

   A. You find capsule-shaped egg cases.

   B. You find scraps of paper and cloth gathered in the corner of a drawer.

   C. You see droppings that look like black grains of pepper.

   D. You smell a strong, oily odor.

13. Which of the following is a sign you might have a problem with cockroaches?

   A. You find small holes burrowed through the storeroom wall.

   B. You find droppings that look like black grains of pepper underneath a refrigeration unit.

   C. You see small piles of sawdust that appear to have fallen from the ceiling.

   D. You find webs and wings in the dry-storage area.

14. All of the following are critical components of an integrated pest management program *except*

   A. denying pests access to the establishment.

   B. denying pests food, water, and a hiding or nesting place.

   C. working with a licensed PCO to eliminate pests that do enter.

   D. notifying the EPA that pesticides are being used in the establishment.

15. When pesticides are applied in the establishment, you must do all of the following *except*

   A. leave stationary equipment uncovered.

   B. remove all movable, food-contact surfaces.

   C. wash, rinse, and sanitize food-contact surfaces that have been sprayed.

   D. make a corresponding MSDS available to employees for the pesticide used.

For answers, please turn to page 11-42.

Apply Your Knowledge   Answers

Page	Activity

11-3   Test Your Food Safety Knowledge

1. False   2. False   3. False   4. False   5. True

11-8   What's Missing?

The following items are missing in the handwashing station: ❶ Soap
❷ Signage indicating that employees must wash hands before returning to
work ❸ A waste container for used paper towels ❹ A warm-air hand dryer.
While this is not always required, it is a good idea in the event that paper
towels run out.

11-17   How Bright Should It Be?

1. C   2. A   3. B   4. C   5. B

11-18   To Sanitize or Not to Sanitize

The following situations require the foodhandler to clean and sanitize the
item they are using: 2, 3, 4, 5

11-30   What's Wrong with This Picture?

The following things are wrong with the three-compartment sink:
❶ There is no clock with a second hand. Employees would not be able
to time how long an item has been immersed in the sanitizer.
❷ Soap suds from the wash sink have been carried over into the
rinse sink and the sanitizer sink. This can deplete the sanitizer.
❸ The temperature of the rinse water is only 90˚F (32˚C). It should be
at least 110˚F (43˚C). ❹ A cleaned and sanitized pot is not being air-
dried properly. It should be inverted.

11-36   Who Am I?

1. R   3. C   5. R   7. C
2. C   4. R   6. C

11-39   Multiple-Choice Study Questions

1. A	4. B	7. C	10. C	13. B
2. D	5. C	8. B	11. C	14. D
3. C	6. C	9. A	12. B	15. A

Apply Your Knowledge   **Notes**

Food Safety Regulation and Standards

Inside this section:
▶ Government Regulatory System for Food
▶ The Food Code
▶ Foodservice Inspection Process

After completing this section, you should be able to:
▶ Identify the principles and procedures needed to comply with food safety regulations.
▶ Identify state and local regulatory agencies and regulations that require food safety compliance.
▶ Prepare for a regulatory inspection.
▶ Identify the proper procedures for guiding a health inspector through the establishment.

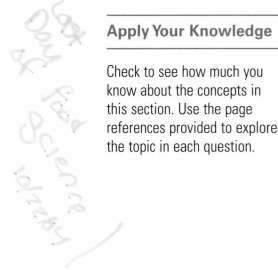

Apply Your Knowledge

Check to see how much you know about the concepts in this section. Use the page references provided to explore the topic in each question.

Test Your Food Safety Knowledge

❶ **True or False:** The FDA writes the food regulations that must be followed by each establishment. *(See page 12-4.)*

❷ **True or False:** Health inspectors are generally employees of the Centers for Disease Control and Prevention (CDC). *(See page 12-4.)*

❸ **True or False:** You should ask to accompany the health inspector during the inspection of your establishment. *(See page 12-6.)*

❹ **True or False:** Critical violations noted during a health inspection should be corrected within one week of the inspection. *(See page 12-6.)*

❺ **True or False:** Establishments can be closed by the health department if they find a significant lack of refrigeration. *(See page 12-7.)*

For answers, please turn to page 12-10.

CONCEPTS

▶ **U.S. Department of Agriculture (USDA):** Federal agency responsible for the inspection and quality grading of meat, meat products, poultry, dairy products, eggs and egg products, and fruit and vegetables shipped across state lines.

▶ **Food and Drug Administration (FDA):** Federal agency that writes the Food Code. The FDA also inspects foodservice operations that cross state borders—interstate establishments such as food manufacturers and processors, and planes and trains—because they overlap the jurisdictions of two or more states. In addition, the FDA shares responsibility with the USDA for inspecting food-processing plants to ensure standards of purity, wholesomeness, and compliance with labeling requirements.

▶ **FDA Food Code:** Recommendations written by the FDA to assist state health departments in developing regulations for a foodservice inspection program.

▶ **Health inspector:** City, county, or state employee who conducts foodservice inspections. Health inspectors are also called sanitarians, health officials, and environmental health specialists. They generally are trained in food safety, sanitation, and public health principles and methods.

INTRODUCTION

There are several reasons why it is important to have a foodservice inspection program. Most important is that failure to ensure food safety can jeopardize the health of your customers and could cost you your business. The purpose of an inspection program is to

▶ evaluate whether the establishment is meeting minimum sanitation and food safety standards.

▶ protect the public's health by requiring establishments to provide food that is safe, uncontaminated, and presented properly.

▶ convey new food safety information to an establishment.

▶ provide an establishment with a written report, noting deficiencies, so it can be brought into compliance with safe food practices.

GOVERNMENT REGULATORY SYSTEM FOR FOOD

Today, government control of food in the U.S. is exercised at three levels: federal, state, and local. At the federal level, the U.S. Department of Agriculture (USDA), and the Food and Drug Administration (FDA), are directly involved in the inspection process.

USDA

▶ The USDA is responsible for inspection and quality grading of meat, meat products, poultry, dairy products, eggs and egg products, and fruit and vegetables shipped across state lines.

U.S. Regulatory System for Food

Federal Level

Regulations recommended

▼

State Level

Regulations written

▼

Local Level

Regulations enforced

Exhibit 12a

▶ The USDA provides these services through the Food Safety and Inspection Service (FSIS) agency.

FDA

▶ The FDA is the agency that writes the Food Code.

▶ The FDA also inspects foodservice operations that cross state borders (interstate establishments such as those on planes and trains, as well as food manufacturers and processors) because they overlap the jurisdictions of two or more states.

▶ The FDA shares responsibility with the USDA for inspecting food-processing plants to ensure standards of purity, wholesomeness, and compliance with labeling requirements.

Regulation

▶ In the U.S., most food regulations affecting restaurant and foodservice operations are written at the state level (except regulations for interstate or international establishments, which are determined at the federal level).

▶ Each state decides whether to adopt the Food Code or some modified form of it. (See *Exhibit 12a.*)

▶ State regulations may be enforced by local (city or county) or state health departments.

▶ City, county, or state health inspectors (also called sanitarians, health officials, or environmental health specialists) conduct foodservice inspections in most states. They generally are trained in food safety, sanitation, and public health principles and methods.

THE FOOD CODE

The Food Code is written by the FDA based upon input from the Conference for Food Protection (CFP). The Food Code lists the government's recommendations for foodservice regulations, which are intended to assist state health departments in developing regulations for a foodservice inspection program. It is not an actual law. Although the FDA recommends adoption by the states, it cannot require it.

Food and sanitation codes are written very broadly and generally cover the following areas:

▶ **Foodhandling and preparation:** sources, receiving, storage, display, service, transportation

▶ **Personnel:** health, personal cleanliness, clothing, hygiene practices

▶ **Equipment and utensils:** materials, design, installation, storage

▶ **Cleaning and sanitizing:** facility, equipment

▶ **Utilities and services:** water, sewage, plumbing, restrooms, waste disposal, integrated pest management (IPM)

▶ **Construction and maintenance:** floors, walls, ceilings, lighting, ventilation, dressing rooms, locker areas, storage areas

▶ **Foodservice units:** mobile, temporary

▶ **Compliance procedures:** foodservice inspections, enforcement actions

It is the responsibility of the manager to keep food safe and wholesome throughout the establishment at all times, regardless of the inspector or the inspection process.

FOODSERVICE INSPECTION PROCESS

Well-managed establishments will perform continuous self-inspections to keep food safe, in addition to the regular inspections performed by the health department. Establishments with high standards for sanitation and food safety consider health department inspections only a supplement to their self-inspection programs.

During health department inspections, the local health code serves as the inspector's guide. It is a good idea to keep a current copy of your local or state sanitation regulations and be familiar with them. Regularly compare the code to procedures at your establishment, but remember that code requirements are only *minimum* standards to keep food safe.

The following suggestions will enable managers and operators to get the most out of food safety inspections:

❶ **Ask for identification.** Clarify the purpose of the visit. Make sure you know whether it is a routine inspection, the result of a customer complaint, or for some other purpose.

❷ **Cooperate.** Answer all of the inspector's questions to the best of your ability. Instruct employees to do the same. Explain to the inspector that you wish to accompany him or her during the inspection. This will encourage open communication and a good working relationship. If a deficiency can be corrected quickly, do so, or tell the inspector when it can be corrected.

❸ **Take notes.** As you accompany the inspector, make a note of any problem pointed out. (See *Exhibit 12b.*) Make it clear you are willing to correct problems. If you believe the inspector is incorrect about something, note what was mentioned. Then ask the inspector's supervisor for a second opinion.

❹ **Keep the relationship professional.** Do not offer food or drink before, during, or after an inspection. This could be viewed as bribery.

❺ **Be prepared to provide records requested by the inspector.** If a request appears inappropriate, you can check with the inspector's supervisor or with your lawyer about limits on confidential information.

❻ **Discuss violations and time frames for correction with the inspector.** Deficiencies and comments should be discussed in detail with the inspector. In order to make complete and permanent corrections, you will need to know the exact nature of the violation, how it impacts food safety, how to correct it, and whether or not the inspector will follow-up.

❼ **Follow up.** Establishments are generally given a short amount of time (forty-eight hours or less) to correct critical violations. Correct the problems. Determine why they occurred by evaluating sanitation procedures, the master cleaning schedule, and employee foodhandling practices. Establish new procedures or revise existing ones to correct the problem permanently, if necessary.

Exhibit 12b

Foodservice Inspection

As you accompany the inspector, make a note of any problem pointed out.

Closure

In some states, if the inspector determines a facility poses an immediate and substantial health hazard to the public, he or she may ask for a voluntary closure, or issue an immediate suspension of the permit to operate. Examples of hazards calling for closure include:

▶ Significant lack of refrigeration

▶ Backup of sewage into the establishment itself or its water supply

▶ Emergency, such as a building fire or flood

▶ Serious infestation of insects or rodents

▶ Long interruption of electrical or water service

SUMMARY

Today, government control regarding food safety in the U.S. is exercised at three levels: federal, state, and local. Recommendations for restaurant and foodservice regulations are written at the federal level by the FDA—in the form of the Food Code. Regulations are written at the state level, and enforcement is usually carried out at the local level. Some agencies at the federal level such as the FDA and the USDA are directly involved in the inspection process.

All establishments must follow standard food safety practices critical to the safety and quality of the food served. An inspection system lets the establishment know how well it is following these practices. Well-managed establishments will perform continuous self-inspections to protect food safety, in addition to the regular inspections performed by the health department. Establishments with high standards for sanitation and food safety consider health department inspections only a supplement to their self-inspection programs.

During the inspection process, cooperate with the health inspector and keep the relationship professional. Tell them that you wish to accompany them during the inspection. When the inspector points out a problem, take notes. If a deficiency can be corrected immediately, do so. If not, tell the inspector when it can be corrected. Be prepared to provide records that are requested. Discuss violations and time frames and then follow up.

Apply Your Knowledge

Multiple-Choice Study Questions

Use these questions to test your knowledge of the concepts presented in this section.

1. An establishment can be closed for all of the following reasons *except*
    A. significant lack of refrigeration.
    B. backup of sewage.
    C. serious infestation of insects or rodents.
    D. minor violations not corrected within twenty-four hours.

2. Which of the following is a goal of the food safety inspection program?
    A. To evaluate whether an establishment is meeting minimum food safety and sanitation standards
    B. To protect the public's health
    C. To convey new food safety information to establishments
    D. All of the above

3. Which operation would most likely be subject to a food safety inspection by a federal agency?
    A. Hospital
    B. Cruise ship crossing international waters
    C. Local ice cream store with a history of safety violations
    D. Food kitchen run by church volunteers

4. A person shows up at a restaurant claiming to be a health inspector. What should the manager do?
    A. Ask to see identification.
    B. Ask to see an inspection warrant.
    C. Ask for a hearing to determine if the inspection is necessary.
    D. Ask for a one-day postponement to prepare for the inspection.

Continued on next page...

5. Which of the following agencies enforces food safety in a restaurant?
   A. FDA
   B. CDC
   C. State or local health department
   D. USDA

6. Violations noted on the health inspection report should be
   A. discussed in detail with the inspector.
   B. corrected within forty-eight hours or less if they are critical.
   C. explored to determine why they occurred.
   D. All of the above

7. The responsibility for keeping food safe in an establishment rests with the
   A. FDA.   C. health inspector.
   B. manager/operator.   D. state health department.

8. Food codes developed by state agencies are
   A. minimum standards necessary to ensure food safety.
   B. maximum standards necessary to ensure food safety.
   C. voluntary guidelines for establishments to follow.
   D. inspection practices for grading meats and meat products.

For answers, please turn to page 12-10.

Apply Your Knowledge   **Answers**

Page	Activity
12-2	Test Your Food Safety Knowledge

1. False
2. False
3. True
4. False
5. True

12-8   Multiple-Choice Study Questions

1. D   5. C
2. D   6. D
3. B   7. B
4. A   8. A

Apply Your Knowledge   Notes

HOW TO IMPLEMENT THE FOOD SAFETY PRACTICES LEARNED IN THE SERVSAFE PROGRAM

Your commitment to food safety does not end when you complete the ServSafe Examination. The ServSafe program has provided you with the essential knowledge to help you keep the food at your establishment safe. Now, it is your responsibility to take that knowledge and implement it in your operation. To do this, you must examine the following aspects of your operation and compare them with your new ServSafe knowledge:

▶ Current food safety policies and procedures

▶ Employee training

▶ Your facilities

The steps listed below will help you implement the practices you have learned in the ServSafe program. The first two steps help you make the comparison that will take you from where you are today to where you need to go to consistently keep food safe in your establishment. Keep in mind that regardless of what your establishment's practices are right now, food safety requires continuous improvement.

❶ **Evaluate your food safety practices using the *Food Safety Evaluation Checklist* on page A-3.** This checklist identifies the most critical food safety practices that must be followed in every operation. It consists of a series of Yes/No questions that will assist you in identifying opportunities for improvement. Wherever a *No* is checked in this evaluation, you have identified a gap in your practices. These gaps will be the starting point for improving your current food safety program.

❷ **Fill out the *Regulatory Requirements Worksheet* on page A-10.** This worksheet will help you identify areas where your local regulatory requirements differ from the information in *ServSafe Essentials*. It is vital that you comply with your local regulatory requirements. Your company policies should be designed to comply with, or exceed, these requirements. This worksheet supplies you with space to perform a direct comparison of what *ServSafe Essentials* states, what the local jurisdiction requires, and what your company policy states regarding each issue. If you do not have a company policy regarding an issue, this is a gap that must be filled.

❸ Determine the causes of the gaps identified in Steps 1 and 2. For example, if your refrigerator is incapable of holding food at 41°F (5°C), this is a gap. There are many things that could have caused this situation, including faulty equipment, an overstocked refrigerator, a refrigerator door that is opened too frequently, etc. You must explore each of these potential causes to determine the true reason for the gap.

❹ Address gaps by creating a solution that may include:

▶ Developing or revising SOPs

▶ Training employees on new or revised SOPs

▶ Implementing SOPs

▶ Bringing existing equipment up to standard or purchasing new equipment

▶ Training or retraining employees

❺ Evaluate your solution periodically to ensure it has addressed the gaps identified in Steps 1 and 2.

FOOD SAFETY EVALUATION CHECKLIST

This checklist focuses on the most critical food safety practices your operation must follow. Its purpose is to identify gaps between your current practices and those practices critical to keeping food safe in an operation. Please take the time to seriously consider each one of these questions. Your job, your customers, and your employees are counting on you!

Directions: Check *Yes* after each question if your establishment currently performs the practice or *No* if it currently does not. Each practice that is checked *No* identifies a gap that offers an opportunity for revising your food safety program. For each question, a page reference has been provided to help you locate the related content in *ServSafe Essentials*. Use these references when revising your food safety program.

Food Safety Evaluation Checklist

Topic/Principle	Evaluation	Page Reference in *Essentials*
Time and Temperature Control		
1. Are calibrated thermometers available to all foodhandlers?	Yes _____ No _____	5-6
2. Do you calibrate thermometers regularly?	Yes _____ No _____	5-6
3. Do you document product temperatures in a temperature log?	Yes _____ No _____	5-6
4. Do employees know how to use thermometers properly?	Yes _____ No _____	5-10 through 5-12
5. Do you minimize the amount of time food spends in the temperature danger zone (41°F [5°C] to 135° F [57°C])?	Yes _____ No _____	5-5
6. Do you reject food that has not been received at the proper temperature?	Yes _____ No _____	6-5 through 6-19
7. Do you store potentially hazardous food at its required storage temperature?	Yes _____ No _____	7-10 through 7-11
8. Do you thaw food properly?	Yes _____ No _____	8-3
9. Do you cook potentially hazardous food to required minimum internal temperatures?	Yes _____ No _____	8-9 through 8-12
10. Do you cool cooked, potentially hazardous food according to the proper time and temperature requirements?	Yes _____ No _____	8-14
11. Do you reheat potentially hazardous food that will be hot held to 165°F (74°C) for fifteen seconds within two hours?	Yes _____ No _____	8-17
12. Do you hold potentially hazardous food at the proper temperature (41°F [5°C] or lower, or 135°F [57°C] or higher)?	Yes _____ No _____	9-3 and 9-4
13. Are time and temperature controls incorporated in your SOPs?	Yes _____ No _____	5-7
14. Are time and temperature controls part of every employee's job?	Yes _____ No _____	5-6

Continued on next page...

A-3

Food Safety Evaluation Checklist *continued*		
Topic/Principle	**Evaluation**	**Page Reference in *Essentials***
Preventing Contamination and Cross-Contamination		
1. Do you store food in a way that prevents contamination?		
a. Do you store food in designated storage areas only?	Yes _____ No _____	7-4
b. Do you store cooked or ready-to-eat food above raw meat, poultry, and fish?	Yes _____ No _____	7-7
c. Do you store food away from walls and at least six inches off the floor?	Yes _____ No _____	7-9
2. Do you prepare food in a way that prevents contamination?		
a. Do you assign specific equipment to each type of food product used in your establishment?	Yes _____ No _____	5-4
b. Do you clean and sanitize all work surfaces, equipment, and utensils after each task?	Yes _____ No _____	5-4
c. When using the same prep table to prepare food, do you prepare raw and ready-to-eat food at different times?	Yes _____ No _____	5-5
d. Do you use ingredients that require minimal preparation?	Yes _____ No _____	5-5
3. Do you hold food in a way that prevents contamination?		
a. Do you cover food to protect it from contaminants?	Yes _____ No _____	9-3
b. Do you discard food being held for service after a predetermined amount of time?	Yes _____ No _____	9-3
4. Do you serve food in a way that prevents contamination?		
a. Do you minimize bare-hand contact with food that is cooked or ready to eat?	Yes_____ No_____	9-5
b. Do servers avoid handling the food-contact surfaces of glassware, dishes, and utensils?	Yes_____ No_____	9-6
c. Do you maintain self-service areas in a way that prevents contamination?	Yes_____ No_____	9-9

Food Safety Evaluation Checklist

Topic/Principle	Evaluation	Page Reference in *Essentials*
Preventing Contamination and Cross-Contamination *continued*		
5. Do you handle chemicals in a way that prevents contamination?		
a. Do you store chemicals away from food, utensils, and equipment?	Yes _____ No _____	3-8
b. Are containers used to dispense chemicals properly labeled?	Yes _____ No _____	3-8
c. If pesticides are applied in the establishment, are all food and food-contact surfaces removed prior to the application?	Yes _____ No _____	11-35
6. Do you only use food-grade utensils and equipment in your establishment?	Yes _____ No _____	3-9
Approved Sources		
1. Do you purchase food from suppliers that obtain their products from approved sources?	Yes _____ No _____	6-3
2. Do you ensure that your suppliers are reputable?	Yes _____ No _____	6-3
3. Do your suppliers deliver during off-peak hours?	Yes _____ No _____	6-3
Proper Personal Hygiene		
1. Are all employees aware of all the ways in which they can contaminate food?	Yes _____ No _____	4-3 through 4-4
2. Do all employees follow the proper procedure for proper handwashing?	Yes _____ No _____	4-6
3. Are all employees aware of the instances when handwashing is required?	Yes _____ No _____	4-7
4. Do all employees follow the proper hand maintenance procedures, such as keeping nails short and clean, and covering cuts and sores?	Yes _____ No _____	4-8
5. Do you provide the right type of gloves in your establishment for handling food?	Yes _____ No _____	4-9

Continued on next page...

Food Safety Evaluation Checklist *continued*

Topic/Principle	Evaluation	Page Reference in *Essentials*
Proper Personal Hygiene *continued*		
6. Do employees change gloves when necessary?	Yes _____ No _____	4-9
7. Do you have requirements in place for proper work attire for foodhandlers?	Yes _____ No _____	4-10
8. Do you require employees to maintain personal cleanliness?	Yes _____ No _____	4-10
9. Do you prohibit employees from smoking, eating, or drinking in food-preparation and warewashing areas?	Yes _____ No _____	4-11
10. Do you have policies in place for handling employee illnesses?	Yes _____ No _____	4-11 through 4-12
Food Safety Systems		
1. Do you have the appropriate prerequisite programs for a food safety system in place?	Yes _____ No _____	10-3
2. Can you identify the five most common CDC risk factors for foodborne illness?	Yes _____ No _____	10-3
3. Do you have a food safety system in place?	Yes _____ No _____	10-4 through 10-10
4. Do you know when a HACCP plan is required?	Yes _____ No _____	10-11 through 10-12
Facilities and Equipment		
1. Are your handwashing stations equipped with the necessary tools and supplies?	Yes _____ No _____	11-7 through 11-8
2. Is the equipment you purchase designed with sanitation in mind?	Yes _____ No _____	11-9
3. Is stationary equipment installed properly?	Yes _____ No _____	11-11
4. Does your equipment receive regular maintenance?	Yes _____ No _____	11-12

Food Safety Evaluation Checklist

Topic/Principle	Evaluation		Page Reference in *Essentials*

Facilities and Equipment *continued*

5.	Is your plumbing installed and maintained by a licensed plumber?	Yes _____	No _____	11-13
6.	Is garbage properly stored and removed from the premises?	Yes _____	No _____	11-17

Cleaning and Sanitizing

1.	Do your employees know the difference between cleaning and sanitizing?	Yes _____	No _____	11-18
2.	Do your employees know how to clean and sanitize food-contact surfaces?	Yes _____	No _____	11-18
3.	Do your employees store cleaning cloths in a sanitizer solution between uses?	Yes _____	No _____	11-24
4.	Do your employees know the frequency for cleaning and sanitizing food-contact surfaces?	Yes _____	No _____	11-18
5.	Do your employees know how to use the sanitizer used in your establishment effectively?	Yes _____	No _____	11-20
6.	Do your dishwashing employees know how to use the warewashing machine properly?	Yes _____	No _____	11-21 through 11-22
7.	Do your dishwashing employees know how to clean and sanitize items in a three-compartment sink properly?	Yes _____	No _____	11-22 through 11-23
8.	Do your employees know how to clean and sanitize stationary equipment properly?	Yes _____	No _____	11-24
9.	Do your employees know how to clean nonfood-contact surfaces?	Yes _____	No _____	11-25
10.	Do your employees know how to store clean and sanitized utensils, tableware, and equipment properly?	Yes _____	No _____	11-26
11.	Do you have a master cleaning program in place?	Yes _____	No _____	11-28 through 11-29

Continued on next page…

Food Safety Evaluation Checklist *continued*

Topic/Principle	Evaluation	Page Reference in *Essentials*
Pest Control		
1. Do you have a contract with a licensed pest control operator?	Yes _____ No _____	11-31
2. Do you inspect deliveries for signs of pests?	Yes _____ No _____	11-31
3. Do you take measures for preventing pests from entering the establishment?	Yes _____ No _____	11-32 through 11-33
4. Do you take measures for denying pests food and shelter in the establishment?	Yes _____ No _____	11-34 through 11-34
5. Can your employees identify signs of pests?	Yes _____ No _____	11-34 through 11-35
6. If used, do you store pesticides properly in the establishment?	Yes _____ No _____	11-36
Auditing (Self-Inspection)		
1. Do you conduct self-inspections?	Yes _____ No _____	12-5
2. Do you regularly compare your local or state sanitation regulations to procedures at your establishment?	Yes _____ No _____	12-5
3. Have you addressed all infractions from your last inspection report?	Yes _____ No _____	12-6
4. Do you have a plan for working with health inspectors during inspections?	Yes _____ No _____	12-6
Employee Training		
1. Do you have food safety training programs for both new and current employees?	Yes _____ No_____	1-6
2. Do you have assessment tools that identify ongoing food safety training needs for employees?	Yes _____ No _____	1-6
3. Do you have food safety training resources that include books, videos, posters, and technology-based materials to meet your employees' learning needs?	Yes _____ No _____	1-6
4. Do you keep records documenting that employees have completed training?	Yes _____ No _____	1-6

Apply Your Knowledge   Notes

REGULATORY REQUIREMENTS WORKSHEET

This worksheet is designed to help you take a closer look at policies and procedures highlighted in *ServSafe Essentials* that might differ from your jurisdictional requirements and/or your company policies.

Directions: Highlighted on the following pages are several food safety issues that need further exploration. For each issue, we have provided a reference in the text. Look at each issue and compare it to your jurisdictional requirements. If the requirements differ from *ServSafe Essentials* or are not explained in detail in the textbook, you should document the requirements accordingly. Contact your regulatory agency if you have any questions. Once you have recorded the requirements, compare it to your company policy on that issue. Use the space provided to document your policy. If you do not have a policy about an issue, then you have identified a gap in your food safety program that must be addressed.

You should also use this worksheet as a tool when talking with your health inspector. Your inspector will provide valuable insight on how to improve food safety in your establishment.

Contact information for your local regulatory agency:

_____

_____

_____

_____

_____

_____

_____

Consumer advisory for consuming raw and undercooked potentially hazardous food (page 1-7)

ServSafe Essentials **states:** In all cases, these high-risk consumers should be advised of any potentially hazardous food (or ingredient) that is raw, or not fully cooked. Tell them to consult a physician before regularly consuming this type of food.

Jurisdictional Requirements

Does your jurisdiction require you to have a consumer advisory?

Yes _____   No _____

If so, the requirements are as follows:

_____

_____

_____

_____

_____

_____

_____

Company Policy

What is your company's consumer advisory policy?

_____

_____

_____

_____

_____

_____

_____

How to wash hands (page 4-6)

***ServSafe Essentials* states:** To ensure proper handwashing in your establishment, train your foodhandlers to follow the steps illustrated in *Exhibit 4a*—Proper Handwashing Procedure.

Jurisdictional Requirements

The jurisdictional requirements for how to wash hands are as follows:

_____

_____

_____

_____

_____

_____

Company Policy

What is your company's procedure for handwashing?

_____

_____

_____

_____

_____

_____

Hand sanitizers (page 4-7)

ServSafe Essentials **states:** Establishments must only use hand sanitizers that have been approved by the FDA.

Jurisdictional Requirements

Are hand sanitizers allowed in your jurisdiction?

Yes _____   No _____

If allowed, the following are approved for use:

_____

_____

_____

_____

_____

_____

Company Policy

What is your company's policy regarding the use of hand sanitizers?

_____

_____

_____

_____

_____

_____

Bare-hand contact with ready-to-eat food (page 4-8)

ServSafe Essentials **states:** For those jurisdictions that allow bare-hand contact with ready-to-eat food, establishments must have a verifiable written policy on handwashing procedures.

Jurisdictional Requirements

Does your jurisdiction allow bare-hand contact with ready-to-eat food?

Yes _____ No _____

If allowed, what are the requirements? If not, what are the requirements?

_____

_____

_____

_____

_____

_____

Company Policy

What is your company's policy regarding bare-hand contact with ready-to-eat food?

_____

_____

_____

_____

_____

_____

False nails and nail polish (page 4-8)

ServSafe Essentials **states:** Some jurisdictions allow single-use gloves to be worn over false or polished nails.

Jurisdictional Requirements

Does your jurisdiction allow foodhandlers to wear false nails?

Yes _____ No _____

If so, under what conditions?

_____

_____

_____

_____

Does your jurisdiction allow foodhandlers to wear nail polish?

Yes _____ No _____

If so, under what conditions?

_____

_____

_____

_____

_____

Company Policy

What is your company's policy regarding false nails and nail polish?

_____

_____

_____

_____

Glove use (page 4-9)

Jurisdictional Requirements

What are the requirements for glove use in your jurisdiction?

_____

_____

_____

_____

Company Policy

What is your company's policy regarding glove use?

_____

_____

_____

_____

Proper work attire (page 4-10)

ServSafe Essentials **states:** A foodhandler's attire plays an important role in the prevention of foodborne illness. Dirty clothes may harbor pathogens and give customers a bad impression of your establishment. Therefore, managers should make sure that foodhandlers observe strict dress standards (i.e., hair restraints, clean uniforms, wearing aprons at the appropriate time, removing jewelry, wearing appropriate shoes, etc.).

Jurisdictional Requirements

The requirements for foodhandler work attire in your jurisdiction are:

_____

_____

_____

_____

Company Policy

What is your company's policy regarding proper work attire?

_____

_____

_____

_____

Drinking in food-preparation and warewashing areas (page 4-11)

ServSafe Essentials **states:** Some jurisdictions allow employees to drink from a covered container with a straw while in food-preparation and warewashing areas.

Jurisdictional Requirements

Does your jurisdiction allow employees to drink in food-preparation and warewashing areas?

Yes _____   No _____

If so, what are the requirements:

_____

_____

_____

_____

Company Policy

What is your company's policy regarding drinking in food-preparation and warewashing areas?

_____

_____

_____

_____

Policies for handling employee illness (page 4-11)

ServSafe Essentials **states:** There are several instances when a foodhandler must either be restricted from working with or around food, or excluded from working within the establishment.

Jurisdictional Requirements

When must employees be restricted from working with or around food?

_____

_____

_____

_____

_____

When should employees be excluded from the establishment?

_____

_____

_____

_____

_____

What is your jurisdiction's reporting requirements for an employee who has been diagnosed with a foodborne illness? What is the procedure for determining when that employee can safely return to work?

_____

_____

_____

_____

_____

Company Policy

When does your company restrict employees from working with or around food?

_____

_____

_____

_____

_____

When does your company exclude employees from the establishment?

_____

_____

_____

_____

_____

What is your company's policy for reporting an employee diagnosed with a foodborne illness? How does your company determine when that employee can safely return to work?

_____

_____

_____

_____

_____

Approved food sources (page 6-3)

ServSafe Essentials **states:** An approved food source is one that has been inspected and is in compliance with applicable local, state, and federal law. Before you accept any deliveries, it is your responsibility to ensure that food you purchase comes from suppliers (distributors) and sources (points of origin) that have been approved.

Company Policy

How do you verify that your suppliers use approved sources?

_____

_____

_____

_____

How do you verify that you are using an approved supplier?

_____

_____

_____

_____

Receiving temperatures (pages 6-5 through 6-19)

ServSafe Essentials **states:** Please refer to *Exhibits 6b* through *6p*.

Jurisdictional Requirements

What are required receiving temperatures for products used in your establishment?

_____

_____

_____

_____

_____

Company Policy

What are the receiving temperatures for products used in your establishment?

_____

_____

_____

_____

_____

Refrigerated storage temperature (page 7-5)

ServSafe Essentials **states:** Refrigerated storage areas are typically used to hold potentially hazardous food at 41°F (5°C) or lower. Some jurisdictions allow food to be held at 45°F (7°C) or lower.

Jurisdictional Requirements

The required refrigerated storage temperature in your jurisdiction is:

_____

_____

_____

_____

_____

Company Policy

The required refrigerated storage temperature in your establishment is:

_____

_____

_____

_____

_____

Preparing and packaging fresh juice (page 8-7)

ServSafe Essentials **states:** If juice is packaged on-site for sale at a later time, the establishment must have a variance from the regulatory agency, and the juice must be treated (i.e., pasteurized) according to an approved HACCP plan or be labeled with the following phrase: *"Warning: This product has not been pasteurized and therefore may contain harmful bacteria that can cause serious illness in children, the elderly, and people with weakened immune systems."*

Jurisdictional Requirements

Do you prepare and package fresh fruit and vegetable juice on-site?

Yes _____   No _____

If so, what does your jurisdiction require?

_____

_____

_____

_____

_____

Company Policy

How do you ensure the safety of the juice that you prepare and package?

_____

_____

_____

_____

_____

Minimum internal cooking temperature for various types of food (pages 8-9 through 8-12)

ServSafe Essentials **states:** Please refer to *Exhibit 8d*—Cooking Requirements for Specific Types of Food—for minimum internal cooking temperatures.

Jurisdictional Requirements

What are the time and temperature cooking requirements in your jurisdiction for the food that you prepare in your establishment?

_____

_____

_____

_____

_____

_____

_____

Company Policy

What are the time and temperature requirements for cooking food that you prepare in your establishment?

_____

_____

_____

_____

_____

_____

_____

Cooling food (page 8-14)

ServSafe Essentials **states:** Cooked food must be cooled from 135°F (57°C) to 70°F (21°C) within two hours and from 70°F (21°C) to 41°F (5°C) or lower in an additional four hours, for a total cooling time of six hours. Some jurisdictions use one-stage cooling, by which food must be cooled to 41°F (5°C) or lower in less than four hours.

Jurisdictional Requirements

What are the time and temperature requirements for cooling food in your jurisdiction?

_____

_____

_____

_____

_____

_____

Company Policy

What are the time and temperature requirements for cooling food in your establishment?

_____

_____

What is the procedure for cooling food in your establishment?

_____

_____

_____

Holding hot and cold potentially hazardous food (pages 9-3 and 9-4)

ServSafe Essentials **states:** Potentially hazardous, cold food must be held at 41°F (5°C) or lower. Potentially hazardous hot food must be held at an internal temperature of 135°F (57°C) or higher. You can also hold it at an even higher temperature of 140°F (60°C) as an additional safeguard.

Jurisdictional Requirements

In your jurisdiction, potentially hazardous, cold food must be held at the following internal temperature:

_____ °F/°C

In your jurisdiction, potentially hazardous, hot food must be held at the following internal temperature:

_____ °F/°C

Company Policy

In your establishment, potentially hazardous, cold food must be held at the following internal temperature:

_____ °F/°C

In your establishment, potentially hazardous, hot food must be held at the following internal temperature:

_____ °F/°C

Holding food without temperature control (page 9-4)

ServSafe Essentials **states:** Ready-to-eat, potentially hazardous food can be displayed or held for consumption without temperature control for up to four hours under the following conditions:

▶ Prior to removing the food from temperature control, it has been held at 41°F (5°C) or lower, or 135°F (57°C) or higher.

▶ The food contains a label that specifies when the item must be discarded.

▶ The food must be sold, served, or discarded within four hours.

Jurisdictional Requirements

Is holding ready-to-eat, potentially hazardous food without temperature control allowed in your jurisdiction?

Yes _____   No _____

If so, what are the requirements?

_____

_____

_____

_____

_____

Company Policy

What is your policy for holding food without temperature control?

_____

_____

_____

_____

_____

When a HACCP plan is required (page 10-11)

ServSafe Essentials **states:** Establishments that perform the following activities must have an approved HACCP plan in place:

▶ Smoke or cure food as a method of food preservation

▶ Use food additives as a method of food preservation

▶ Package food using a reduced-oxygen packaging method

▶ Offer live, molluscan shellfish from a display tank

▶ Custom-process animals for personal use

▶ Package unpasteurized juice for sale to the consumer without a warning label

Jurisdictional Requirements

What additional activities require an approved HACCP plan?

_____

_____

_____

_____

_____

Approval of layout and design plans (page 11-6)

ServSafe Essentials **states:** Many jurisdictions require approval of layout and design plans by the local regulatory agency.

Jurisdictional Requirements

Does your jurisdiction require layout and design approval?

Yes _____   No _____

What agencies must approve it?

_____

_____

_____

Handwashing station (page 11-7)

ServSafe Essentials states: Refer to *Exhibit 11a.*

Jurisdictional Requirements

What are the requirements for a handwashing station in your jurisdiction?

_____

_____

_____

_____

_____

_____

_____

Company Policy

What are your requirements for handwashing stations?

_____

_____

_____

_____

_____

_____

_____

Private water source (page 11-12)

ServSafe Essentials **states:** If your establishment uses a private water supply, such as a well, rather than an approved public source, check with your local regulatory agency for information on inspections, testing, and other requirements.

Jurisdictional Requirements

Does your water come from a private source, such as a well?

Yes _____   No _____

What are the inspection and testing requirements?

_____

_____

_____

_____

_____

_____

_____

Company Policy

What are your inspection and testing requirements?

_____

_____

_____

_____

_____

_____

_____

Securing safe water during an disruption (page 11-12)

ServSafe Essentials **states:** If the water supply is disrupted, follow these guidelines to continue serving food safely:

▶ Use bottled water.

▶ Boil water.

▶ Purchase ice.

▶ Use boiled water for essential cleaning, such as pots and pans. Consider using single-use items (plates and utensils) to minimize warewashing. Keep a supply of boiled, warm water available for handwashing.

Jurisdictional Requirements

How does your jurisdiction notify your establishment when the water supply has become unsafe?

_____

_____

_____

_____

What are the approved sources for water during an emergency?

_____

_____

_____

_____

Company Policy

What procedures does your company follow when your water supply is disrupted?

_____

_____

_____

_____

Apply Your Knowledge

Notes

Index

Index

W

Y